10 YEARS OF FOCUS THERE IS A MONOGRAPH

名贵珠宝鉴赏与投资
——便携式资产配置

10年专注 才有专著

陈小雨 / 主编

作者介绍
Introduction of the Author

陈小雨 博士，国际宝石学院特聘教师，英国FGA国际珠宝鉴定师，多家上市公司和家族企业的名贵珠宝资产配置顾问。积累了10年名贵珠宝理论以及市场实战经验，多次深入斯里兰卡、泰国、缅甸宝石矿区考察，多次参加日本、迪拜、美国和中国香港地区等地的国际珠宝展，专注于名贵珠宝的鉴赏与资产配置。

Dr. Chen. Xiaoyu is a distinguished teacher at the International Gemological Institute, an international jewelry appraiser of FGA in the United Kingdom, and an asset allocation consultant for many listed companies and family businesses. She has accumulated 10 years of experience in luxury jewelry theory and market practice. She has visited gem mining areas in Sri Lanka, Thailand and Myanmar for many times, and participated in international jewelry exhibitions in Japan, Dubai, the United States and Hong Kong, etc., focusing on the appreciation and asset allocation of luxury jewelry.

本书内容简介
Content Introduction

不论你是珠宝爱好者还是投资者，如果你想了解或者投资名贵珠宝，我相信这本书肯定不会让你走弯路。这本书分为上篇和下篇，上篇讲的是12种名贵珠宝的鉴赏，包括红宝石、蓝宝石等，它的颜色、证书、重量等是如何影响珠宝的价格的，以及近10年每种珠宝的价格趋势图表。下篇讲的是名贵珠宝的投资逻辑，包括我在美国华尔街财富管理公司对于名贵珠宝投资的调研，包括我参加佳士得、苏富比、保利拍卖的拍卖调研，这本书很系统地介绍了名贵珠宝投资的投资历史、发展情况、以及投资风险等。

Whether you are a jewelry lover or an investor, if you want to get to know or invest in luxury jewelry, I believe this book will certainly not let you take a roundabout course. This book is divided into two parts, including Part one and Part two. Part one is about the appreciation of 12 kinds of luxury jewelry, including ruby, sapphire, etc., and about how the color, certificate, weight, etc., affects the price of the jewelry. And I concluded the price trend chart of each jewelry in the past 10 years. Part two is about the investment logic of luxury jewelry, including my research on luxury jewelry investment in Wall Street wealth management company in the United States, and also including my participation in Christie's, Sotheby's and Poly Auction. This book systematically introduces the investment history, development, and investment risks of luxury jewelry investment.

Preface 序言

写这本书的起因，是一个富二代的珠宝投资的赚钱经验刺激了我。我是在跟她一起去参加国外的珠宝访学认识的，我那个时候还在上学，她已经大学毕业10年了。她是个富二代，没有上过一天班，这10年她的主要工作就是花钱。我本来正想鄙视她的时候，却被她赚钱的思维上了一课。她在英国读书的时候就喜欢珠宝，她不仅自己喜欢也影响了她妈妈身边的朋友，慢慢地，她就变成了买珠宝的代购，读书4年做了3年代购。她发现，越是名贵的好珠宝，每年的拿货的成本价涨得越多，超过了她之前卖出的价格。于是她跟家里人商量，她大学毕业以后就只买高品质的名贵珠宝，但是不马上卖，至少放个3年、5年，甚至10年，再慢慢卖出。她做对了，果然投资名贵珠宝就是做时间的朋友，她在2014年花30万元买的10克拉的哥伦比亚祖母绿，当时真的也是高价了，但是2023年这颗祖母绿在市面上的价格是300万元了，她以低于市场价20%，也就是240万元卖出，很快就被同行珠宝商买走了。2014年的30万元变成了2023年的240万元，这个投资回报让她爸很满意，比她爸创业的回报还高，哈哈哈。

回归主题。大家好，我是小雨老师，也是一名名贵珠宝投资顾问，为多家上市公司和家族企业做名贵珠宝的咨询服务。从本科到博士我花了10年时间，一边做理论研究一边走遍世界各个宝石矿区做市场实践，才写出了这本《名贵珠宝鉴赏与投资》。不论你是珠宝爱好者还是投资者，如果你想了解或者投资名贵珠宝，我相信这本书肯定不会让你走弯路。这本书分为上篇和下篇，上篇讲的是12种名贵珠宝的鉴赏，包括红宝石、蓝宝石等，它的颜色、证书、重量等是如何影响珠宝的价格的，以及近10年每种珠宝的价格趋势图表。下篇讲的是名贵珠宝的投资逻辑，包括我在美国华尔街财富管理公司对于名贵珠宝投资的调研，包括我参加佳士得、苏富比、保利拍卖的拍卖调研，这本书很系统地介绍了名贵珠宝投资的投资历史、发展情况、以及投资风险等。我不仅写了书，还推出线上和线下的名贵珠宝鉴赏与投资的课程，这个课程会慢慢进入各大商学院和金融机构。小雨老师真心希望能够帮助大家在鉴赏和投资名贵珠宝上都不走弯路。

与其说这是一本关于名贵珠宝鉴赏与投资的圣经，不如说这是我10年间在国内外宝石矿区工作学习的宝石交易心得。我之前在美院学艺术，而后去清华学了设计，后来去泰国学宝石，再后来去美国哈佛研究名贵珠宝的资产配置，反正一系列的读书都围绕着珠宝这个行业。参加过国内外的一些大型的珠宝展，比如说中东迪拜和中国香港地区珠宝展，泰国、斯里兰卡、日本、美国等地的国际珠宝展。这些知名的大型国际珠宝展，让我自己学到了很多的知识，跟全球的珠宝商、珠宝矿主都有深度的交流合作。不论是理论还是市场实践，都用了10年时间，写这本书就是为了纪念自己10年的青春。

最后，关于这本书，我要感谢很可爱的人。第一要感谢清华博士毕业的田卫老师，他提供了很多法律和金融方面的支持。第二要感谢我的助理邢萱萱女士，她毕业于中国地质大学珠宝专业，她提供了图片的支持。第三要感谢四川外语大学的龙兴春教授，他提供了翻译的支持。第四要感谢清华美院的师姐、西南民族大学的教授刘春燕教授，她提供了书籍排版的支持。第五要感谢同行业的珠宝人，他们提供了宝石图片的支持。第六要感谢我的管家王丽娟阿姨，在无数个深夜写书的时候，她每天做营养餐保证我的健康。第七要感谢我自己，10年专注，才有了专著。

Preface

The reason for writing this book is that I was stimulated by the profitable experience of a rich friend's jewelry investment. I got to know her when we went on a jewelry visit abroad. I was still in school at that time, and she had graduated from college for 10 years. She was born with a silver spoon in her mouth and has never done a day's work. In these 10 years her main job was to spend money. I was about to despise her, but she taught me a lesson in how to make money. When she studied in the UK, she liked jewelry. She not only liked it herself, but also influenced her mother's friends. Gradually, she became a buyer of jewelry. Within the four years of studying, she was a jewelry buyer for three years. She found that the purchase price of the more expensive pieces rose more each year, surpassing the price she had previously sold them for. So she discussed with her family that after she graduated from college, she would only buy high-quality luxury jewelry, but not sell it immediately. She would wait for at least for three, five, or even 10 years, and then slowly sell it. She did the right thing, proving that investing in luxury jewelry is to be a friend of time. She spent 300,000 yuan in 2014 to buy 10 carat Colombian emerald, which was really high. But in 2023 this emerald in the market price is 3 million yuan, and she sold at 20% below the market price, that is, 2.4 million yuan. It was soon purchased by a fellow jeweler.300,000 yuan in 2014 became 2.4 million yuan in 2023, the return on investment made her father very satisfied, which is higher than the return of her father's business, how interesting!

Let's back to the main topic. Hello everyone, I am Xiaoyu, and also a luxury jewelry investment consultant, providing consulting services for many listed companies and family business. From undergraduate to doctoral, I spent 10 years doing theoretical research while traveling around the world to do market practice in gem mining areas before writing this book <Luxury Jewelry Appreciation and Investment>. Whether you are a jewelry lover or an investor, if you want to understand or invest in fine jewelry, I believe this book will certainly not let you take a roundabout course. This book is divided into two parts, including Part one and Part two. Part one is about the appreciation of 12 kinds of luxury jewelry, including ruby, sapphire, etc., and about how the color, certificate, weight, etc., affects the price of the jewelry. And I concluded the price trend chart of each jewelry in the past 10 years. Part two is about the investment logic of luxury jewelry, including my research on luxury jewelry investment in Wall Street wealth management company in the United States, and also including my participation in Christie's, Sotheby's and Poly Auction. This book systematically introduces the investment history, development, and investment risks of luxury jewelry investment. Not only have I written a book, but I have also launched online and offline courses on the appreciation and investment of luxury jewelry, which will slowly be introduced into major business schools and financial institutions. Ms. Xiaoyu sincerely hopes to help everyone avoid making detours in the appreciation and investment of luxury jewelry.

This book is rather as my experience in gem trading during my 10 years of working and learning in gem mining areas at home and abroad, rather than a bible on the appreciation and investment of luxury jewelry. I studied art at the Academy of Fine Arts, then went to Tsinghua University to study design, then went to Thailand to study gemstones, and then went to Harvard in the United States to study the asset allocation of luxury jewelry. Anyway, that was a series of studies that focused on the jewelry industry. I had participated in some large jewelry exhibitions at home and abroad, such as the Middle East Dubai and Hong Kong jewelry exhibitions, Thailand, Sri Lanka, Japan, the United States and other international jewelry exhibitions. These well-known large-scale international jewelry exhibitions have allowed me to learn a lot of knowledge, and have in-depth exchanges and cooperation with jewelers and jewelry mine owners around the world. Whether it is theory or market practice, it took 10 years to write this book to commemorate my 10 years of youth.

Finally, for this book, I want to thank some very lovely people. First, I would like to thank Tian Wei, a PhD graduate from Tsinghua University, who has provided a lot of legal and financial support. Second, I would like to thank my assistant Ms. Xuanxuan Xing, who graduated from the jewelry major of China University of Geosciences, who provided the support of the pictures. Third, I would like to thank Professor Long Xingchun of Sichuan Foreign Languages University for his translation support. Fourth, I would like to thank Professor Liu Chunyan, an older sister of Tsinghua Academy of Fine Arts and professor of Southwest University for Nationalities, who provided support for book typesetting. Fifth, I would like to thank the jewelers in the same industry who have provided the support of the gemstone pictures. Sixth, I want to thank my housekeeper, Aunt Wang Lijuan, who makes nutritious meals every day to ensure my health while writing books at countless late nights. Seventh, I want to thank myself, 10 years of focus, just have a monograph.

Introduction
导论

小雨老师在美国纽约47街珠宝考察

2023年，我去美国纽约的华尔街拜访一些财富管理公司，发现在很多富豪的资产配置项目里，有一个项目叫做另类投资，其中就含有名贵珠宝这个类别，这是富人圈子里默认的理财方式。他们的理财顾问跟我分享了一个理论，他说股票、基金、期货等资产配置都是人为构建起来的金融服务系统，也有可能因为人为的原因被打破，虽然这样的概率很小，但是有这样的可能。所以富人在资产配置的时候，会考虑一些稀缺又能流通的资产，比如名贵珠宝。名贵珠宝是大自然几十亿年的结晶，不会因为人为的干预瞬间消失，即便这个世界因为人祸混乱了，这些稀缺的名贵珠宝或许也会变成交易的筹码，这就是为什么富人会配置名贵珠宝的原因，不仅仅是漂亮，更多的是价值的流通。

In 2023, I went to the Wall Street in New York to visit some wealth management companies. And I found that in the asset allocation projects of many rich people, there was a project called alternative investment, which contains the category of luxury jewelry, which was the default way of financing in the rich circle. Their financial advisor shared a theory with me. He said that asset allocation such as stocks, funds and future goods were all artificially constructed financial service systems, which may be broken due to artificial reasons. Although the probability is very small, it is possible. Therefore, when the rich allocate assets, they will consider some scarce and negotiable assets, such as luxury jewelry. Luxury jewelry is the crystallization of nature for billions of years, which will not disappear instantaneously because of human intervention, even if the world because of man-made chaos, the scarce luxury jewelry may also become a trading weight, which is why the rich will configure luxury jewelry. It's not only because of its beauty, but also the value of the circulation.

After

后来我又去了全世界最出名的名贵珠宝交易中心——美国纽约47街考察，这里有来自全世界的珠宝商人以及想卖珠宝的客人。我在47街拜访了几个犹太人，他们跟我说纽约47街其实是"二战"以后才开始形成的一条珠宝交易商业街。"二战"的时候，犹太人被纳粹一路驱赶追杀，逃到了美国纽约47街避难，当时他们身上没有其他可以交换食物的东西，只带了一些名贵的珠宝，因为珠宝比较小、比较好隐藏，他们逃到了纽约47街以后，就把珠宝放到一些砖缝里藏起来，饿了就靠珠宝来交换食物。

后来慢慢发展，纽约47街就变成了全世界著名的珠宝交易中心。让人很奇怪的是，纽约47街到处挂着BUYER的字样，这里的商人其实不是靠卖珠宝赚钱，他们是靠买珠宝赚钱。不论你来自于哪里，你只要拿出很名贵的珠宝，这里的人就会出价，你就可以卖掉换成金钱。你还可以到处对比价格，名贵的珠宝在这里可以卖一个好价格。我接下来要跟大家普及一下哪些是可以作为资产配置的名贵珠宝。

我们的珠宝大概可以分成四类：一是国际五大珠宝，之所以叫做国际五大宝石，是因为这5种宝石价值比较名贵，产量比较稀有，主要是指红宝石、蓝宝石、祖母绿、猫眼、钻石；一是半宝石，之所以叫做半宝石，主要是其价格相对比较便宜，产量比较大，主要是指碧玺、坦桑石、海蓝宝等；一是玉石，主要是指翡翠、和田玉等；一是有机珠宝，主要是指珍珠、珊瑚、海螺珠等。这只是一个基本的分类，并不能完全说明价值。比如半宝石包括了尖晶石，大多数尖晶石价值不高，但是高品质的马亨盖尖晶石价值非常高，也是拍卖场上的新星，非常具有保值的潜力。而国际五大珠宝里面的钻石，比较有保值能力的是彩色钻石，即便小克拉的彩色钻石也比较有保值的能力，但是白色钻石，只有超级大克拉的，才有保值的能力。

After that, I went to the 47th Street in New York, the most famous jewelry trading center in the world, where there were jewelry merchants from all over the world and customers who want to sell jewelry. I visited several Jews on 47th Street, and they told me that in fact, 47th Street in New York became a jewelry trading street starting since the end of World War II. In the Second World War, Jews were chased by the Nazis all the way to the 47th Street in New York in the United States. At that time, they had no other things to exchange for food, only with some luxury jewelry, as jewelry was relatively small and better to hide. When they fled to New York 47th Street, they put the jewelry in some bricks to hide, and used them to exchange for food when they were hungry.

Later, the 47th Street in New York became a famous jewelry trading center in the world. It is quite strange to see the BUYER signs all over 47th Street in New York, where merchants don't actually make their money by selling jewelry. Instead, they make their money by buying jewelry. It doesn't matter where you are from, as long as you bring out very valuable jewelry, people here will bid on it, and you can sell it for money. You can also compare prices everywhere, and luxury jewelry can sell for a good price here. Now, I am going to introduce to you about what kind of luxury jewelry can be used as an asset allocation.

Our jewelry can be divided into four categories in general. The first one is the international Five major jewelry, the reason why it is called the international five major gems, is that the value of these five gems are relatively high and the production is relatively small. This kind of jewelry category mainly refers to ruby, sapphire, emerald, cat's eye, diamond. The second one is semi-luxury gem, the reason why it is called semi-luxury gem, mainly because its price is relatively cheap, and the production is relatively large. This kind of jewelry mainly refers to tourmaline, tanzanite, aquamarine and so on. The third one is jade jewelry, mainly refers to jade, Hetian jade, etc. The fourth one is the organic jewelry, mainly refers to pearls, coral, conch beads and so on. This is only a very basic classification and does not fully representing their value. For example, semi-luxury gems, including spinel, most of their value is not high. But those high-quality Mahengai spinel, are having very high value. And they are also the new star in the auction, very potential to preserve value. Meanwhile, the diamond in the international five major jewelry, the colored diamonds are more possible to perverse value, even for those small carat colored diamonds. On the other hand, white diamonds, only super large carat, can preserve value.

所以珠宝的分类只是一个相对的概念，有一些半宝石的价值可能比五大国际珠宝的价值还高，不能通过这个分类一概而论珠宝的价值。

Therefore, the classification of jewelry is only a relative concept, and there are cases that the value of some semi-luxury stones may be higher than the value of five international jewelry. Hence, the value of jewelry cannot be generalized through this classification.

小雨老师在美国纽约跟珠宝商交流

小雨老师在哈佛大学上课

本书主要讲的是名贵珠宝，所谓的名贵珠宝主要是指价值比较高。有一些来自国际五大珠宝，有些来自半宝石，有些来自有机珠宝，唯独不包含翡翠。翡翠在中国人的圈子里有流通的价值，而在国际上的流通价值比较低，认可度比较低，变现比较难，除非是个别极品的翡翠珠串，国外有藏家会收藏。而其他名贵珠宝在国际上流通比较好，认可度比较高，在美国、英国、瑞士，都是可以在市场上流通的，变现能力比较强。本书讲的名贵珠宝，也是在国际拍卖场上比较受欢迎的12种珠宝，包括：蓝宝石、红宝石、尖晶石、帕帕拉恰蓝宝石、帕拉伊巴碧玺、祖母绿、亚历山大变石、金绿猫眼、黑欧泊、彩色钻石、海螺珠、野生珍珠。

This book is mainly about luxury jewelry, and the so-called luxury jewelry mainly refers to their relatively high value. Some of them are from the international Five jewels, some from semi-luxury gems, some from organic jewelry, but no jade included. Jade in the Chinese circle has the value of circulation, but in the international circulation, jade's value is relatively low. Jade's recognition is relatively low, and it is more difficult for realization. Unless it is an individual jade beads, and there are collectors abroad will collect. And other luxury jewelry are better circulated in the international market, with relatively high recognition. In the United States, the United Kingdom, Switzerland, these jewelries can be circulated well in the market, and the cash ability is relatively strong. The luxury jewelry mentioned in this book are also the 12 kinds of jewelry that are more popular in the international auction market, including: sapphire, ruby, spinel, Papalacha sapphire, Paraiba tourmaline, emerald, Alexandrite, golden green cat's eye, black Opal, colored diamonds, conch beads, wild pearls.

目录
CONTENT

| 序言 Preface | 001 |
| 导论 Introduction | 001 |

上篇：名贵珠宝鉴赏篇
Part One: Luxury Jewelry Appreciation

第一章 蓝宝石
01 Chapter One Sapphire

一、蓝宝石的基本特征 —— 002
Basic Characteristics of Sapphire
二、如何鉴赏和投资收藏蓝宝石 —— 004
How to Appreciate and Invest in the Sapphire Collection
（一）蓝宝石的鉴赏 —— 005
Appreciation of Sapphire
（二）蓝宝石的投资收藏建议 —— 017
Sapphire Investment and Collection Suggestions

第二章 红宝石
02 Chapter Two Ruby

一、红宝石的基本特征 —— 024
Basic Characteristics of Ruby
二、如何鉴赏和投资收藏红宝石 —— 026
How to Appreciate and Invest in the Ruby Collection
（一）红宝石的鉴赏 —— 027
Appreciation of Rubies
（二）红宝石的投资收藏建议 —— 040
Rubies Investment and Collection Suggestions

第三章 尖晶石
03 Chapter Three Spinel

一、尖晶石的基本特征 —— 046
Basic Characteristics of Spinel
二、如何鉴赏和投资收藏尖晶石 —— 048
How to Appreciate and Invest in the Spinel Collection
（一）尖晶石的鉴赏 —— 049
Appreciation of Spinel
（二）尖晶石的投资收藏建议 —— 057
Spinel Investment and Collection Suggestions

Content

第四章 帕帕拉恰
Chapter Four Padparadscha

一、帕帕拉恰的基本性质 —— 064
Basic Characteristics of Padparadscha

二、如何鉴赏和投资收藏帕帕拉恰 —— 066
How to Appreciate and Invest in the Padparadscha Collection

（一）帕帕拉恰的鉴赏 —— 067
Appreciation of Padparadscha

（二）帕帕拉恰的投资收藏建议 —— 073
Padparadscha Investment and Collection Suggestions

第五章 帕拉伊巴
Chapter Five Paraiba

一、帕拉伊巴的基本特征 —— 078
Basic Characteristics of Paraiba

二、如何鉴赏和投资收藏帕拉伊巴 —— 080
How to Appreciate and Invest in the Paraiba Collection

（一）帕拉伊巴的鉴赏 —— 081
Appreciation of Paraiba

（二）帕拉伊巴的投资收藏建议 —— 089
Paraiba Investment and Collection Suggestions

第六章 祖母绿
Chapter Six Emerald

一．祖母绿的基本特征 —— 096
Basic Characteristics of Emerald

二．如何鉴赏和投资收藏祖母绿 —— 098
How to Appreciate and Invest in the Emerald Collection

（一）祖母绿的鉴赏 —— 099
Appreciation of Emerald

（二）祖母绿的投资收藏建议 —— 108
Emerald Investment and Collection Suggestions

第七章 亚历山大变石
Chapter Seven Alexandrite

一、亚历山大变石的基本特征 —— 114
Basic Characteristics of Alexandrite

二、如何鉴赏和投资收藏亚历山大变石 —— 116
How to Appreciate and Invest in the Alexandrite Collection

（一）亚历山大变石的鉴赏 —— 117
Appreciation of Alexandrite

（二）亚历山大变石的投资收藏建议 —— 123
Alexandrite Investment and Collection Suggestions

Content

第八章 金绿猫眼
08 Chapter Eight
Golden Green Cat's Eye

一、金绿猫眼的基本特征 —— 128
Basic Characteristics of Golden Green Cat's Eye

二、如何鉴赏和投资收藏金绿猫眼 —— 130
How to Appreciate and Invest in the Golden Green Cat's Eye Collection

（一）金绿猫眼的鉴赏 —— 131
Appreciation of Golden Green Cat's Eye

（二）金绿猫眼的投资收藏建议 —— 136
Golden Green Cat's Eye Investment and Collection Suggestions

第九章 黑欧泊
09 Chapter Nine
Black Opal

一、黑欧泊的基本特征 —— 140
Basic Characteristics of Black Opal

二、如何鉴赏和投资收藏黑欧泊 —— 142
How to Appreciate and Invest in the Black Opal Collection

（一）黑欧泊的鉴赏 —— 142
Appreciation of Black Opal

（二）黑欧泊的投资收藏建议 —— 144
Black Opal Investment and Collection Suggestions

第十章 彩色钻石
10 Chapter Ten
Colored Diamond

一、彩色钻石的基本特征 —— 148
Basic Characteristics of Colored Diamonds

二、如何鉴赏和投资收藏彩色钻石 —— 150
How to Appreciate and Invest in the Colored Diamond Collection

（一）彩色钻石的鉴赏 —— 151
Appreciation of Colored Diamonds

（二）彩色钻石的投资收藏建议 —— 157
Colored Diamond Investment and Collection Suggestions

第十一章 海螺珠
11 Chapter Eleven
Conch Pearl

一、海螺珠的基本特征 —— 162
Basic Characteristics of Conch Pearl

二、如何鉴赏和投资收藏海螺珠 —— 164
How to Appreciate and Invest in the Conch Pearl Collection

（一）海螺珠的鉴赏 —— 165
Appreciation of Conch Pearls

（二）海螺珠的投资收藏建议 —— 169
Conch Pearl Investment and Collection Suggestions

CONTENT

第十二章 野生珍珠
12 Chapter Twelve
Natural Pearl

一．野生珍珠的基本特征 —————————— 174
Basic Characteristics of Natural Pearls
二．如何鉴赏和投资收藏野生珍珠 ————— 177
How to Appreciate and Invest in the Natural Pearl Collection
（一）野生珍珠的鉴赏 ——————————— 178
Appreciation of Natural Pearl
（二）野生珍珠的投资收藏建议 ——————— 180
Natural Pearl Investment and Collection Suggestions

下篇：名贵宝石投资篇
Part Two: Luxury Jewelry Investment

第一章 名贵珠宝作为投资资产
01 Chapter One
Luxury Jewelry as Investment Asset

一、名贵珠宝投资的历史和发展 ——————— 184
The History and Development of Luxury Jewelry Investment
二、名贵珠宝在资产配置中的时机和作用 ——— 187
The Timing and Role of Luxury Jewelry in Asset Allocation
三、名贵珠宝投资市场及其吸引力 —————— 191
Luxury Jewelry Investment Market and Its Attractiveness
四、名贵珠宝投资与高净值家族财富管理 ——— 196
Luxury Jewelry Investment and High Net-value Family Wealth Management

Content

第二章 名贵珠宝投资与其他投资的对比分析
02 Chapter Two
Comparative Analysis of Luxury Jewelry Investment and Other Investments

一、名贵珠宝投资的比较优势 —————————— 201
The Comparative Advantage of Luxury Jewelry Investment

二、名贵珠宝投资与黄金投资的比较分析 —————— 203
Comparative Analysis of Luxury Jewelry Investment and Gold Investment

三、名贵珠宝投资与股市投资的主要区别 —————— 206
The Main Differences Between Luxury Jewelry Investment and Stock Market Investment

四、名贵珠宝投资与房地产投资的主要区别 ———— 208
The Main Differences Between Luxury Jewelry Investment and Real Estate Investment

五、名贵珠宝投资与基金投资的主要区别 —————— 211
The Main Differences Between Luxury Jewelry Investment and Fund Investment

六、名贵珠宝投资与理财投资的主要区别 —————— 213
The Main Differences between Luxury Jewelry Investment and Financial Investment

七、名贵珠宝投资与投资理财型保险的主要区别 —— 216
The Main Differences Between Luxury Jewelry Investment and Financial Insurance Investment

第三章 名贵珠宝投资的市场趋势
03 Chapter Three
The Market Trend of Luxury Jewelry Investment

一、全球主要名贵珠宝市场和交易中心 ——————— 220
The World's Major Luxury Jewelry Market and Trading Center

二、当前国内名贵珠宝市场特点 ——————————— 223
The Current Domestic Luxury Jewelry Market Characteristics

三、全球和区域市场特点 ——————————————— 224
Global and Regional Market Characteristics

四、可持续名贵珠宝投资的重要性 —————————— 225
The Importance of Sustainable Jewelry Investment

五、案例研究：名贵珠宝投资案例 —————————— 227
Case Study: Luxury Jewelry Investment Case

第四章 名贵珠宝的供需关系、价格与价值
04 Chapter Four
The Supply and Demand Relationship, Price and Value of Luxury Jewelry

一、名贵珠宝的保值与增值潜力 ——————————— 232
The Preservation and Appreciation Potential of Luxury Jewelry

二、名贵珠宝供需动态及其对价格的影响 —————— 237
The Supply and Demand Dynamics of Luxury Jewelry and Its Impact on Prices

三、名贵珠宝价格趋势、市场动态与投资时机 ———— 240
Luxury Jewelry Price Trends, Market Dynamics and Investment Opportunities

四、影响名贵珠宝价值的因素 ———————————— 243
Factors that Influence the Value of Luxury Jewelry

第五章 名贵珠宝投资策略与风险管理
05 Chapter Five
Luxury Jewelry Investment Strategy and Risk Management

一、名贵珠宝投资的短期与长期策略 —— 248
Short-term and Long-term Strategies for Luxury Jewelry Investment

二、投资组合构建策略 —— 250
Investment Portfolio Construction Strategy

三、名贵珠宝投资的风险评估 —— 255
Risk Assessment of Luxury Jewelry Investment

四、名贵珠宝投资的风险管理 —— 258
Risk Management of Luxury Jewelry Investment

第六章 名贵珠宝投资顾问服务
06 Chapter Six
Luxury jewelry Investment Advisory Services

一、名贵珠宝投资顾问服务中的核心内容 —— 262
The Core Content of Luxury Jewelry Investment Advisory Services

二、名贵珠宝投资顾问服务中的客户画像 —— 264
Portraits of Clients in Luxury Jewelry Investment Advisory Services

三、名贵珠宝投资顾问服务中的财务规划 —— 265
Financial Planning in Luxury Jewelry Investment Advisory Services

第七章 附录：名贵珠宝投资资源
07 Chapter Seven
Appendix: Luxury Jewelry Investment Resources

一、主要的名贵珠宝评估和鉴定机构名录 —— 270
List of Major Luxury Jewelry Evaluation and Appraisal Institutions

二、名贵珠宝展览、交易会和拍卖信息 —— 275
Luxury Jewelry Exhibition, Trade Fair and Auction Information

三、其他珠宝鉴赏 —— 280
Appreciation of Other Jewelries

上篇：名贵珠宝鉴赏篇
Part One: Luxury Jewelry Appreciation

第一章
01 Chapter One

蓝宝石
Sapphire

一、蓝宝石的基本特征
Basic Characteristics of Sapphire

蓝宝石的英文名称为 Sapphire，源于拉丁文 Sapphirus，意为蓝色。是国际流通得五大宝石之一，在国际市场上非常好流通，也是拍卖场上非常受欢迎的名贵珠宝之一，很多国际知名珠宝大品牌都会用到蓝宝石。蓝宝石是刚玉宝石中除红宝石之外的其他颜色刚玉的通称。除了红宝石外，自然界中的蓝色、淡蓝色、绿色、黄色、粉红色等各种颜色的宝石级刚玉都被称为蓝宝石。蓝色的蓝宝石是最受欢迎的彩色宝石之一，并且在古埃及、古希腊和古罗马时期就已经被王室成员佩戴和使用。

The name Sapphire is derived from the Latin word Sapphirus, meaning blue. It is one of the five major gems in international circulation, which is very good in circulation in the international market. It is also one of the most popular luxury jewelry on the auction house, and many internationally renowned jewelry brands are using sapphire. Sapphire is the general name for corundum of other colors except ruby in corundum gems. In addition to rubies, jewel-grade corundum of various colors such as blue, light blue, green, yellow, and pink in nature are all called sapphire. Blue sapphires are one of the most popular colored gemstones and have been worn and used by royal families since ancient Egypt, Greece and Rome.

Basic Features of Sapphire

蓝宝石切割

泰国蓝宝石矿区交易市场

蓝宝石买卖

斯里兰卡蓝宝石矿区交易市场

Basic Characteristics of Sapphire

The blue color of sapphire is one of the most popular colored gemstones and has been worn and used by royals since ancient Egypt, ancient Greece and Rome.

二、如何鉴赏和投资收藏蓝宝石
How to Appreciate and Invest in Sapphire Collection

  我们主要从以下7个方面去鉴赏和投资一颗蓝宝石，我在每一个鉴赏指标分析里都给出了目前的市场情况以及相应指标的价格影响，也参考了一些拍卖价格，大家可以参考。

  We mainly appreciate and invest a sapphire from the following 7 aspects. In each appreciation index analysis, I have given the current market situation and the price impact of the corresponding indicators, and also referred to some auction prices for your reference.

（一）蓝宝石的鉴赏
Appreciation of Sapphire

1. 颜色
Color

是指蓝宝石有哪些颜色。
It refers to the color of the sapphire.

蓝色蓝宝石
Blue Sapphire

吉尔德实验室将蓝宝石的颜色分为浅、中、浓、艳、深等级。其中，"皇家蓝"和"矢车菊蓝"是对蓝色蓝宝石最高的颜色评价。"矢车菊蓝"是一种略带紫色调的蓝色，给人朦胧的天鹅绒般质感。而"皇家蓝"是一种正蓝或略带紫色调的蓝色。10克拉的高品质皇家蓝蓝宝石，目前的价格在200万元人民币左右，10克拉的高品质矢车菊蓝蓝宝石价格大概在150万元人民币左右。

皇家蓝蓝宝石

The Gilder laboratory divides the color of sapphire into light, medium, rich, abundant and deep grades. Among them, "royal blue" and "cornflower blue" are the highest color evaluation of blue sapphire. " Cornflower blue " is a slightly purplish shade of blue that gives a hazy velvety texture. And "royal blue" is a normal blue or purplish shade of blue. The current price of 10 carat high-quality royal blue sapphire is about 2 million yuan, and the price of 10 carat high-quality cornflower blue sapphire is about 1.5 million yuan.

粉色蓝宝石
Pink Sapphire

如果是粉色蓝宝石，我们尽量选颜色浓郁一点的粉色，或者热粉色，价值会更高一些，粉色蓝宝石是可以上拍卖会的品类，实物非常漂亮，很少女心，很多大牌像宝格丽、尚美等国际著名品牌都会用到粉色蓝宝石。10克拉的高品质无烧粉色蓝宝石价格也要大几十万元人民币。

If it is pink sapphire, we try to choose a strong color of pink, or hot pink, because the value will be higher. Pink sapphire belongs to the category that can be on the auction house. The physical item is very beautiful, very young, many big brands like Bulgari, Beauty and other international famous brands are using pink sapphire. 10 carat high-quality non-burning pink sapphire price is also hundreds of thousands of yuan.

粉色蓝宝石

紫色蓝宝石
Purple Sapphire

紫色蓝宝石，我们尽量选颜色浓郁一点的紫色，证书上标明VVD Purple，商业名称也叫薰衣草紫色。一颗10克拉的高品质紫色蓝宝石价格也要几十万元人民币。

For purple sapphire, we try to choose a strong color of purple, whose certificate indicated VVD Purple, and the business name is also called lavender purple. A 10 carat high-quality purple sapphire can cost hundreds of thousands of yuan.

紫色蓝宝石

白色蓝宝石 White Sapphire	黄色蓝宝石 Yellow Sapphire	绿色蓝宝石 Green Sapphire	橙色蓝宝石 Orange Sapphire

白色蓝宝石	黄色蓝宝石	绿色蓝宝石	橙色蓝宝石

总结 Sum Up

我们在看蓝宝石各种颜色的时候，都可以按照以下两点来评估。第一，看它的颜色的浓郁程度，浓郁一些的往往要比颜色浅一点的贵，但是浓郁到发黑也是不可取的。蓝色系里面最有价值的两种颜色就是皇家蓝和矢车菊蓝。第二，看颜色的均匀程度，尽量避免一边颜色浓郁一边颜色暗淡。颜色非常均匀的肯定要比颜色不均匀的贵。

When we look at various colors of sapphire, we can evaluate it according to the following two points. First, look at the intensity of its color. The richer color ones are often more expensive than the lighter color one, but if it's too thick to look loke black is not desirable. The two most valuable colors in the blue family are royal blue and cornflower blue. Second, look at the uniformity of the color, try to avoid that whose one side is rich color while the other side is dim color. An even colored sapphire is definitely more expensive than an uneven colored one.

小雨老师经验分享 Experience Sharing from Ms. Xiaoyu

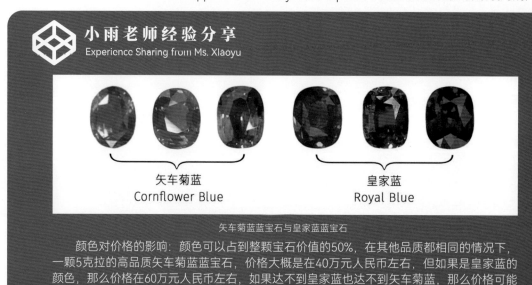

矢车菊蓝蓝宝石与皇家蓝蓝宝石

颜色对价格的影响：颜色可以占到整颗宝石价值的50%，在其他品质都相同的情况下，一颗5克拉的高品质矢车菊蓝蓝宝石，价格大概是在40万元人民币左右，但如果是皇家蓝的颜色，那么价格在60万元人民币左右，如果达不到皇家蓝也达不到矢车菊蓝，那么价格可能要降低一半。一般情况下皇家蓝要比矢车菊蓝贵一些，但具体品质还要具体分析。

The impact of color on price: Color can account for 50% of the value of the whole stone. In the case of other qualities are the same, a 5 carat high-quality cornflower blue stone, its price is about 400,000 yuan, but if it is the color of royal blue, then the price is about 600,000 yuan. If it is not royal blue or cornflower blue, then the price may be reduced by half. Under normal circumstances, the royal blue is more expensive than the cornflower blue, but specific quality will also need specific analysis.

2.净度 Clarity

是指蓝宝石的内外的纯净程度，比如内部是否含有包裹体，外部是否有矿缺，或者开放性的裂隙。蓝宝石最多的净度问题就是，大多数蓝宝石都有色带，这是蓝宝石天然生长的特征，如果没有影响到美观，我们都可以接受。还有一个就是蓝宝石的针状包裹体，也是蓝宝石的天然特性，不影响整体美观的话，我们也是可以接受的。

This refers to the degree of purity inside and outside the sapphire, such as whether the internal contains inclusions, whether there is a external ore shortage, or open cracks. The most clear problem of sapphire is that most sapphire has a ribbon, which is a characteristic of natural growth of sapphire. If it does not affect the beauty, that is acceptable. There is also a needle-like inclusion of sapphire, which is also a natural characteristic of sapphire, and it is acceptable if it does not affect the overall beauty.

蓝宝石瑕疵度

× 延伸至表面的裂隙　　× 明显矿缺　　× 明显的黑包体

1 包裹体的大小、数量和位置
Size, Quantity and Location of Inclusions

包裹体越大、数量越多，对宝石的净度影响就越大。特别是当包裹体位于台面部位时，对宝石净度的影响更为显著。
The larger and more inclusions, the greater the impact on the clarity of the stone. Especially when the inclusion is located in the mesa, the influence on the clarity of the gem is more significant.

2 未愈合的裂隙
An Open Fissure

如果宝石内部有未愈合的裂隙，将降低宝石的耐久性，容易受到损伤。
If there are unhealed cracks inside the stone, the durability of the stone will be reduced and it will be vulnerable to damage.

3 包裹体
Inclusion

少量包裹体存在于宝石内部不会影响宝石的净度质量，比如这个包裹体是透明的，或者包裹体的颜色跟宝石本身的颜色很相似，其实并不会太影响宝石的价值。
The presence of a small amount of inclusion in the gem does not affect the clarity quality of the stone, such as the inclusion is transparent, or the color of the inclusion is very similar to the color of the gem itself. In fact, that does not affect the value of the stone.

蓝宝石戒指

在选择蓝宝石时，确实可以尽量选择表面没有矿坑、明显色带、裂隙和深色包裹体的宝石。大多数蓝宝石都有色带，只要没有太影响美观，都是可以接受的。原则上，只要肉眼难以看到蓝宝石内部特征，即可认为宝石具有良好的净度。并不需要追求完全干净，因为完全干净的蓝宝石反而需要怀疑宝石是否为合成的，如确认为天然生成，则价格非常地贵。

In the selection of sapphire, it is indeed possible to choose gems without pits, obvious color bands, cracks and dark inclusions on the surface. Most sapphires have a ribbon, which is acceptable as long as it does not affect the appearance too much. In principle, as long as it is difficult to see the internal characteristics of sapphire with the naked eye, the stone can be considered to have good clarity. It is not necessary to pursue completely clean, because completely clean sapphire needs to doubt whether the stone is synthetic, if it is confirmed that it is naturally generated, the price will be very expensive.

小雨老师经验分享
Experience Sharing from Ms. Xiaoyu

蓝宝石的净度对价格的影响远远没有蓝宝石的颜色对价格的影响大，如果是克拉数小的还好，我们可以要求净度高一点，但是克拉数越大，我们对净度的要求可能就不需要那么高。从肉眼或者社交距离看，没有影响它的美观即可。拍卖场上一颗15克拉的蓝宝石，它的净度可能没有你想象的好，比如说它有矿缺，有小的愈合裂隙，但是也不会太影响它整体的拍卖价格。在市面上，一颗特别干净的10克拉高品质蓝宝石，价格可能是200万元左右，如果净度差一点，那么它的价格可能就是160万元。

The impact of the clarity of sapphire on the price is far less than the impact of the color of sapphire on the price. If it is a small carat, there will not be a problem, and we can ask for a higher clarity. But the larger carats, our requirements for clarity may not need to be so high. From the naked eye or social distance, there is no impact on its beauty, it is acceptable. A 15-carat sapphire on the auction house may not be as clear as you think, such as it has a lack of ore, or there are small healing cracks. But it will not affect its overall auction price. In the market, a particularly clean 10-carat high-quality sapphire, the price may be about 2 million yuan, and if the clarity is not so good, then its price may be 1.6 million yuan.

3. 切工
Cut

是指蓝宝石外表切割的对称度和工整度。
It refers to the symmetry and evenness of the sapphire surface cutting.

切工　　　　　　切工问题

　　切工主要表现在，对称性和抛光质量等。对称性的问题指正侧面轮廓的对称性偏差、台面偏心、底尖偏心、亭部膨胀、刻面畸形和刻面尖点不尖等。抛光质量指的是宝石表面的抛光程度。

　　The cutting is mainly manifested in symmetry and polishing quality. The symmetry problems include the symmetry deviation of the side profile, the eccentricity of the mesa, the eccentricity of the bottom tip, the bulge of the pavilion, the facet deformity and the non-cusp of the facet. Polishing quality refers to the degree of polishing of the gem's surface.

小雨老师经验分享
Experience Sharing from Ms. Xiaoyu

蓝宝石制作

　　切工对价格的影响：蓝宝石的切工主要是考虑一个空窗问题，也就是说台面过大，它的厚度就会过薄，就会形成一个空窗效应，就是中间那一块颜色很淡，或者说中间看不到颜色，失去了蓝色的浓郁感。还有就是考虑切割过厚的问题，如果台面的比例过小，整颗宝石就会特别厚，色彩会有一点点暗，整比例过小，整颗宝石就会特别厚，色彩会有一点点暗，整个火彩没那么闪亮。切割过厚会涉及价值问题，如果这是一颗5克拉的蓝宝石，但是由于它的台面小，看着像3克拉，你就会觉得不划算，如果切得过厚，我们可以将宝石重新切割。相反如果这颗宝石切割很薄，已经有了空窗效应，我们是没法将宝石重新切割的，只能通过设计镶嵌 尽量弥补。

　　The impact of cutting on the price: the cutting of sapphire is mainly to consider an empty window problem, that is, the mesa is too large, its thickness will be too thin, then it will form an empty window effect. In other word, the middle of the color is very light, or even can not be seen, making it lose the rich sense of blue. There is also the problem of cutting too thick, the proportion of the mesa is too small, the whole gem will be particularly thick. And the color will be a little dark, the whole fire color is not so shiny. If it's a 5 carat sapphire, but with the small mesa, it looks like only a 3 carats. you won't think it's a good deal. And if it's too thick, we can cut the stone again. On the contrary, if the stone is cut very thin and has a void effect, we can not re cut the stone, but can only make up for it by designing the setting.

小雨老师经验分享
Experience Sharing from Ms. Xiaoyu

　　重量对价格的影响：重量对蓝宝石价格的影响挺大的，如果只是克拉数小的，比如说1到3克拉的，重量对价格的影响没有那么大。但如果是5克拉、10克拉，甚至20克拉的，价格影响就特别大。举个例子，比如说3克拉、4克拉，价格都是三四十万元这个样子。但是一旦上了10克拉的高品质蓝宝石，这个蓝宝石的价格可能就要200万元左右，那么到了20克拉，那可能价格要800万元。

　　The effect of weight on price: The impact of weight on the price of sapphire is very large, if it is only a small carats, such as 1 to 3 carats, the impact of weight on the price is not so great. But if it is 5, 10, or even 20 carats, the price impact will be particularly large. For example, for example, for 3 carats or 4 carats, the price is three or four hundred thousand yuan. But for the 10 carat high-quality sapphire, the price of this sapphire may be about 2 million yuan. And then for 20 carat sapphire, it may be 8 million yuan.

---

4. 克拉重量
Carat weight

　　1到4克拉大小的蓝宝石，还是相对多的，上5克拉的高品质蓝宝石就非常少了，上10克拉的高品质蓝宝石非常稀少，价格会成倍增长，尺寸对于高品质蓝宝石的价格影响更加显著。

　　For 1 to 4 carat size sapphire, the quantity is relatively large. But for more that 5 carat high-quality sapphire, it is very few. And for 10 carat high-quality sapphire, that is very rare. The price will increase exponentially, meaning that there is a significant impact of size on the price of high-quality sapphire.

重量

5.产地
Place of Origin

是指蓝宝石的出产地，对价格影响挺大的。
It refers to the origin of sapphire, which has a great impact on the price.

克什米尔
Kashmir

克什米尔蓝宝石戒指

克什米尔蓝宝石被认为是最佳商业品级蓝宝石的代名词，因其产量稀少而闻名。克什米尔蓝宝石有矢车菊蓝色和皇家蓝色，矢车菊蓝明亮鲜艳，带有天鹅绒般的光泽。皇家蓝是浓郁的蓝色调，有时会带有一点点紫色调。克什米尔已经绝矿了，大重量的克什米尔蓝宝石，只存在一些宝石投资者手里，有时在一些大型国际拍卖场上会看见。

Kashmir sapphire is considered synonymous with the best commercial grade sapphire and is known for its rare production. Kashmiri sapphires come in cornflower blue and royal blue. Cornflower blue is bright and vivid with a velvety luster. Royal blue is a rich blue tone, sometimes with a hint of purple. Kashmir has been exhausted, and the large carat Kashmir sapphire is only found in the hands of gemstone investors, and sometimes in some large international auction houses.

斯里兰卡
Sri Lankan

斯里兰卡蓝宝石

斯里兰卡蓝宝石在市场上比较常见，具有丰富的颜色和高透明度。除了蓝色外，斯里兰卡还产出帕帕拉恰蓝宝石，这是一种呈现粉橙或橙红色的珍贵品种。

Sri Lankan sapphires are relatively common in the market, with rich colors and high transparency. In addition to blue, Sri Lanka also produces Papalacha sapphires, a prized variety with a pink orange or orange-red color.

缅甸
Burma

缅甸蓝宝石

缅甸蓝宝石以其神秘感和高贵光环而闻名。它具有深邃的蓝色，无论在何种光照下都不会改变色调，但不具备天鹅绒光泽。缅甸蓝宝石在抹谷、孟素等矿区产出。

Burmese sapphires are known for their mystery and noble aura. It has a deep blue color that does not change its tone in any kind of light, but does not have a velvet sheen. Myanmar sapphire is produced in Mogok, Mengsu and other mining areas.

马达加斯加
Madagascar

马达加斯加蓝宝石

马达加斯加也会产出一些高品质蓝宝石，其市场大约70%由斯里兰卡人控制。马达加斯加蓝宝石有时候会有一些灰色调，价格就会比较便宜。

Madagascar also produces some high-quality sapphires, and its market is about 70% controlled by Sri Lankans. Madagascar sapphires sometimes have a gray tinge, which makes them cheaper.

总结
Sum up

蓝宝石的产地对它的价格影响非常大，目前绝矿了的产地是克什米尔，会在一些古董首饰上看到克什米尔蓝宝石，价格也是最高的，然后是缅甸蓝宝石，再就是斯里兰卡跟马达加斯加的蓝宝石。从市面上的产量来讲，斯里兰卡和马达加斯加的蓝宝石总量要比缅甸和克什米尔多很多。

The origin of sapphire has a very big impact on its price. At present, the production mine in Kashmir has been exhausted. If you see Kashmir sapphire in some antique jewelry, the price is the highest. Then comes to Myanmar sapphire, and then Sri Lanka and Madagascar sapphire. In terms of market production, the total amount of sapphire in Sri Lanka and Madagascar is much more than that in Myanmar and Kashmir.

小雨老师经验分享
Experience Sharing from Ms. Xiaoyu

产地对价格的影响：单从产地这一个因素出发，克什米尔最贵，其次是缅甸，然后是斯里兰卡和马达加斯加。这几个产区的高品质蓝宝石，以他们的价格为例打个比方，一颗克什米尔的3克拉蓝宝石可能需要200万元，一颗缅甸的3克拉蓝宝石可能要中几十万元，一颗斯里兰卡的3克拉蓝宝石能要小几十万元，马达加斯加的蓝宝石跟斯里兰卡的蓝宝石价格差不多。目前从产地上他们的价格分布，基本上是这样一个排名。

The effect of origin on price: Based on origin impact alone, Kashmir is the most expensive, followed by Myanmar, then Sri Lanka and Madagascar. For example, a 3-carat sapphire in Kashmir may cost 2 million yuan, a 3-carat sapphire in Myanmar may cost hundreds of thousands of yuan, a 3-carat sapphire in Sri Lanka can be also hundreds of thousands of yuan but less. And the price of sapphire in Madagascar is similar to that in Sri Lanka. At present, this is basically the ranking status of the price distribution of sapphires according to their origin place.

8.17ct蓝宝石

6.优化
Optimization

是指这颗蓝宝石是否天然，是否经过优化处理
It refers to whether the sapphire is natural, whether it has been optimized.

热处理
Heat Treatment

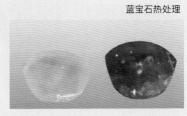
蓝宝石热处理

国际上非常认可的优化方式。热处理又叫加热，是对蓝宝石最常见的一种优化处理方法，通过高温的加热和冷却过程，可以改变蓝宝石的颜色、净度、透明度和光学效应。热处理可以削弱深色蓝宝石的棕色，增强蓝宝石的蓝色。

Internationally recognized optimization methods. Heat treatment, also known as heating, is one of the most common optimization methods for sapphire. Through the high-temperature heating and cooling process, you can change the color, clarity, transparency and optical effects of sapphire. Heat treatment can weaken the brown color of dark sapphire and enhance the blue color of sapphire.

辐射处理
Irradiation Treatment

同一颗蓝宝石在紫外辐照下的颜色变化

国际上非常不认可的处理方式。辐照处理是将蓝宝石暴露于人造辐射源下，以改变其颜色。辐照处理可以使蓝宝石呈现出不同的颜色，但其中一些颜色可能是不稳定的，在光线下会褪色。

This is a highly un-recognized approach internationally. Irradiation is the process of exposing a sapphire to an artificial source of radiation in order to change its color. Irradiation can make sapphires take on different colors, but some of these colors can be unstable and fade under light.

扩散处理
Diffusion Treatment

   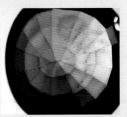

扩散处理蓝宝石　　　　　　　　蓝宝石扩散处理　　　　　蓝宝石扩散处理2

国际上非常不认可的处理方式。扩散处理是在高温条件下，将致色化合物加入蓝宝石，以改善宝石的颜色。扩散处理可以发生在蓝宝石的表面或内部，表面扩散处理可以改变宝石的颜色，但颜色往往不均匀，进行扩散处理可以使颜色渗透到整颗宝石。

This is a highly un-recognized approach internationally. Diffusion treatment is the addition of chromogenic compounds to sapphire under high temperature conditions to improve the color of the stone. Diffusion treatment can occur on the surface or inside of the sapphire. Surface diffusion treatment can change the color of the gem, but the color is often uneven, and diffusion treatment can make the color penetrate into the entire gem.

填充处理
Filling Treatment

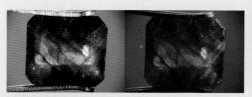

钴玻璃充填蓝宝石

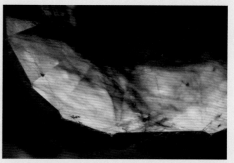

钴玻璃填充蓝宝石中不均匀网状颜色分布及气泡

　　国际上非常不认可的处理方式。填充处理是将一些物质（如有机玻璃）填充到蓝宝石的裂隙中，以改善宝石的亮度、透明度和颜色。填充处理会在蓝宝石中留下特征，如光泽差异、残留气泡和观察时的闪光效应。

　　This is a highly un-recognized approach internationally. Filling treatment is to fill some substance (such as plexiglass) into the cracks of the sapphire to improve the brightness, transparency and color of the stone. The filling treatment leaves characteristics in the sapphire such as gloss differences, residual air bubbles, and a flash effect when viewed.

总结
Sum up

　　蓝宝石分为无加热的、加热的和处理的这三种情况，而无加热的跟加热的都是天然的蓝宝石，因为加热是属于优化，是国际上认可的。只有处理的才是国际上不认可的，它没有市场价值。市面上大多数蓝宝石都有加热。在欧美市场，客户不介意是否加热过。不是每颗蓝宝石经过加热就可以变得很好，只有少部分蓝宝石加热以后会变得更漂亮，所以一颗加热的漂亮的蓝宝石也是值得购买的。但是辐照处理、扩散处理、填充处理是国际上不认可的，并且在证书上也会标明这是处理过的。

加热处理蓝宝石证书

　　Sapphire is divided into three cases: unheated, heated and treated. Unheated and heated are natural sapphire, because heating is optimized and is internationally recognized. Only those being treated are not recognized internationally as it has no market value. Most sapphires on the market are heated. In the European and American markets, customers do not mind if it is heated. Not every sapphire can become very good after heating, only a small part of the sapphire will become more beautiful after heating, so a beautiful sapphire is also worth buying. However, irradiation treatment, diffusion treatment, filling treatment is not recognized internationally, and it will be marked on the certificate that it is treated.

 小雨老师经验分享
Experience Sharing from Ms. Xiaoyu

　　无加热与加热、处理对蓝宝石价格的影响：一颗10克拉的高品质的斯里兰卡的天然无加热的蓝宝石，价格大概是200万元。一颗10克拉斯里兰卡的加热蓝宝石目前市场上的价格也需要中几十万元，而如果是一颗处理的10克拉的斯里兰卡蓝宝石，在市场上是没有价值的。

蓝宝石戒指

　　The impact of heating and heating and treatment on the price of sapphire: for a 10 carat high-quality natural unheated sapphire from Sri Lanka, the price is about 2 million yuan. A 10-carat Sri Lankan heated sapphire currently on the market price also costs hundreds of thousands of yuan, and if it is a processed 10-carat Sri Lankan sapphire, there is no value in the market.

7. 证书
Certificates

不同的证书代表不同的含金量。
Different Certificates Represent Different Gold Content.

01 Gubelin

古柏林（Gübelin）实验室是一家在宝石鉴定和研究领域享有高度声誉的瑞士实验室，在拍卖场上经常出现，证书含金量非常高。如果宝石品质足够好，也会有专门的铂金证书。古柏林会对宝石的重量、产地、有无加热、颜色做分级。尤其是颜色这一栏，古柏林很少给到皇家蓝，他们对皇家蓝色调的要求很高。古柏林会对这颗宝石有一个综合的评分，如果评分在80分以上就是很不错的宝石，如果在90分以上是精品。

The Gubelin Laboratory, a Swiss laboratory with a high reputation in the field of gemstone identification and research, is often seen at auction, and the certificate has a very high gold content. If the stone quality is good enough, there will also be a special platinum certificate. Gubelin grades the stones for weight, origin, heat and color. Especially in the column of color, Gubelin rarely gives royal blue, and they have high requirements for royal blue tones. Gubelin will have a comprehensive score for this gem, if the score is above 80, it is a very good gem, and if it is above 90, that is a very fine product.

02 SSEF
Swiss Gemmological Institute

来自瑞士，非常权威的国际证书，在拍卖场上经常出现，含金量非常高。会对宝石的重量、产地、有无加热、颜色做分级。但是颜色能给到皇家蓝的情况也是极少的

A very authoritative international certificate from Switzerland, it is often found at auction, with a very high gold content. The stones are graded for weight, origin, heating, and color. However, the color can be give as the royal blue is rare.

古柏林证书

SSEF证书

03 AGL
American Gemological Laboratories

AGL证书

　　来自美国，是非常严苛的一家宝石实验室，在欧美市场的含金量很高，也会出现在拍卖场上。会对宝石的重量、产地、有无加热、颜色做分级。但是它没有皇家蓝这样商业化的名称，它会有一个整体的打分，比如颜色占到多少分。通过分数大家也可以评估出这颗宝石的质量。

　　From the United States, it is a very strict gem laboratory, the gold content in the European and American markets is very high, also appeared in the auction. The stones are graded for weight, origin, heating, and color. But it doesn't have a commercial name like Royal Blue, and it has an overall score, like how many points the color accounts for. You can also evaluate the quality of the stone by the score.

04 GRS

GRS 皇家蓝蓝宝石

　　来自瑞士的宝石实验室，主要做彩色宝石的鉴定，含金量也很高，也会出现在拍卖场上，如果宝石品质足够好，还会有铂金证书。GRS会标明产地、有无加热以及颜色的评级。在GRS实验室最新的标准中，皇家蓝的范围增加了intense to vivid blue（鲜蓝至艳蓝）和vivid to deep blue（艳蓝至深蓝）两类。达到矢车菊蓝颜色标准的蓝宝石，GRS则会在颜色一栏备注blue（GRS-type"cornflower"）。在GRS实验室评价标准中，皇家蓝一般要达到vivid blue级别，在颜色一栏会标注 vivid blue（GRS-type"royal blue"）。没有达到矢车菊蓝和皇家蓝级别的蓝色蓝宝石则只备注blue。

　　A gemstone laboratory in Switzerland, it mainly does the identification of colored gems, the gold content is also high, also appeared in the auction. If the gemstone quality is good enough, there will be a platinum certificate. GRS will indicate the origin, the presence or absence of heating, and the color rating. In the latest GRS laboratory standard, the range of royal blue has been added to the intense to vivid blue (bright blue to brilliant blue) and vivid to deep blue (abundant blue to deep blue). For sapphires that meet the cornflower blue color standard, GRS will note Blue in the color column (GRS-type "cornflower"). In the GRS laboratory evaluation criteria, the royal blue is generally meaning achieving the vivid blue level, and the color column will be marked vivid blue (GRS-type "Royal Blue"). Blue sapphires that do not reach the cornflower and Royal Blue levels are only noted blue.

05 GIA
Gemological Institute of America

GIA证书

来自美国，主要做钻石，也可以做彩色宝石的鉴定，提供全面的宝石分析，包括是否经过加热优化、提供产地证明，但是对于颜色很少会给到皇家蓝，只有特别好的蓝宝石才会给到皇家蓝，所以颜色分级这块在GIA的证书上体现得没有那么商业化。

From the United States, it mainly does diamonds identification, and also the identification of colored gems. It provide a comprehensive gem analysis, including whether it has been optimized by heating, provide proof of origin. But for the color, gems will rarely be given to the royal blue, only particularly good sapphire will be given to the royal blue, so the color grading in the GIA certificate is not so commercial.

06 AIGS
Asian Institute of Gemological Sciences

AIGS证书

亚洲宝石学院是一家总部位于泰国曼谷的宝石学院。专注红蓝宝石的分级，包括是否经过加热优化、提供产地证明、颜色分级。

The Asian Gemological Institute is a gemological institute headquartered in Bangkok, Thailand. It focus on the classification of red sapphire, including whether it is optimized by heating, providing proof of origin, color classification.

015

07 GUILD

来自美国,主要做祖母绿的证书,也有其他彩宝的证书。GUILD也会标注产地、有无加热、颜色。在吉尔德实验室的评价标准中,矢车菊蓝属于Vivid blue和Intense blue的一部分;皇家蓝是一种正蓝或略带紫色调的蓝色,属于Vivid blue和Deep blue的一部分。

From the United States, it mainly does the emerald certificate, and there are also other jewerly certificate. GUILD also notes origin, heating, and color. In Gilder's lab, the blue was part of a spectrum of Vivid blue and Intense blue; Royal blue is a positive or purplish shade of blue that is part of Vivid blue and Deep blue.

GUILD 皇家蓝蓝宝石证书

08 EGL

EGL皇家蓝蓝宝石

来自斯里兰卡,在斯里兰卡当地比较出名,主要做蓝宝石的证书。包括是否经过加热优化、提供产地证明、颜色分级。

From Sri Lanka, more famous in local Sri Lanka, mainly does sapphire certificate, including whether it has been optimized for heating, providing proof of origin, and color grading.

小雨老师经验分享
Experience Sharing from Ms. Xiaoyu

蓝宝石的证书对价格的影响:证书分三个梯队:第一梯队的证书包括古柏林证书,SSEF证书,AGL证书。一般情况下很名贵的蓝宝石,或者上拍卖会的蓝宝石,我们会出具这三个证书,在国际上收藏级别的蓝宝石认可度是非常高的。第二梯队的证书主要是指GRS证书,对于精品的蓝宝石,一张GRS证书已经足够了。第三梯队的证书,主要是指AIGS,适合大众宝石,出证书的效率也非常高,很受市场欢迎。以上鉴定机构都是行业内公认的历史比较久远、口碑也比较好的机构。但是还是要说明一下,证书只是一个看宝石的辅助资料,这颗宝石真正的品质还是主要看宝石本身,我们买的也是这颗宝石本身的价值,而不是这张证书。有些宝石只配了一张普通的证书,但是也有可能是一颗很好的宝石;有些宝石配了一张很好的证书,但不一定是很好的宝石。

Sapphire certificate impact on the price: the certificate is divided into three echelons. The first echelon of certificates including Gubelin certificate, SSEF certificate, AGL certificate. Under normal circumstances, very expensive sapphire, or sapphire to be auctioned, we will issue these three certificates, as recognition of the collection level of sapphire is very high in the international market. The second tier of certificates, mainly refers to the GRS certificate. For boutique sapphire, a GRS certificate is enough. The third tier of certificates, mainly referred to as AIGS, is suitable for public gems, and the efficiency of the certificate is also very high, which is very popular in the market. The above appraisal institutions are recognized in the industry with a long history and a good reputation. However, it is still necessary to explain that the certificate is only a supplementary information to judge the gemstone. The real quality of the gemstone is mainly to see the gemstone itself. We buy the value of the gemstone itself, not the certificate. Some stones come with just a plain certificate, but it can also be a very good stone; Some gemstones come with a good certificate, but not necessarily is a very good gemstone.

（二）蓝宝石的投资收藏建议
Sapphire Investment and Collection Suggestions

1.蓝宝石涨价的原因
Reasons for the Price Rise of Sapphire

(1) 产量低
Low Production

小雨老师在斯里兰卡宝石矿区

我在斯里兰卡宝石矿区工作的时候，发现斯里兰卡人开采蓝宝石的工具极其简单，就是纯人工开采，政府不让用器械开采。由于气候、政治、地理条例等综合因素，蓝宝石每年的开采量越来越少，很多矿区都面临着绝矿。2015年我去一个矿主家考察蓝宝石，那个时候他们家每年的矿产量是300克拉，但是2023年的矿产量只有50克拉。即便开采出蓝宝石，能够达到皇家蓝蓝宝石品质的也很稀少，但是皇家蓝蓝宝石真的非常受皇室和富人的喜欢，导致高品质的无烧皇家蓝蓝宝石和矢车菊蓝蓝宝石价格从2010年的5克拉4万元人民币，涨到2023年5克拉40万元人民币。能够达到保值增值的蓝宝石估计只能占到蓝宝石总数的千分之一。大多数商场里的蓝宝石首饰都不会起到保值增值的作用，因为品牌的溢价比较高，而蓝宝石首饰在进行二次销售的时候，主要根据主石的品质给出相应的价格。

When I was working in the gem mining area of Sri Lanka, I found that the Sri Lankan sapphire mining tools are extremely simple, which is pure manual mining, and the government does not allow the use of equipment mining. Due to climate, politics, geographical regulations and other comprehensive factors, sapphire mining is less and less every year, and many mining areas are faced with no ore. In 2015, I went to a mine owner's home to investigate sapphire. At that time, their annual mine production was 300 carats, but the mine production in 2023 was only 50 carats. Even if the sapphire is mined, it is rare to reach the quality of the royal blue sapphire. But the royal blue sapphire is really very popular with the royal family and the rich, resulting in the price of high-quality, non-burned royal blue gem and cornflower blue gem of 5 carat increased from 40,000 yuan in 2010, to 400,000 yuan in 2023. It is estimated that the amount of sapphire that can preserve its value, can only account for one thousand of the total number. Most of the sapphire jewelry in the mall will not play a role in preserving and increasing value, because the brand premium is relatively high, and when the sapphire jewelry is sold again, the corresponding price is mainly given according to the quality of the main gem.

(2) 需求量大
Large Demand

高品质的蓝宝石在发达国家的一线城市认可度是非常高的，尤其是美国、英国、瑞士、阿联酋和迪拜等，名贵珠宝的交易非常发达。中国在近5年需求量也是稳步提升，主要客户来自北京、上海、广州、深圳等发达城市。

high-quality sapphire is very high in the first-tier cities of developed countries, especially in the United States, the United Kingdom, Switzerland and Dubai, the United Arab Emirates, the trading of precious jewelry is very developed. China's demand in the past five years is also steadily increasing, the main customers from Beijing, Shanghai, Guangzhou, Shenzhen and other developed cities.

2.高品质蓝宝石近几年的大约价格涨幅
The Approximate Price Increase of High-Quality Sapphire in Recent Years

蓝宝石近5年价格涨幅
Sapphire Price Increase in the Past 5 Years

重量（克拉） Weight（Carat）	5.00克拉(元人民币/克拉) 5.00ct（CNY/ct）		10.00克拉(元人民币/克拉) 10.00ct（CNY/ct）		20.00克拉(元人民币/克拉) 20.00ct（CNY /ct）	
处理 Treatment	未经热处理 None	热处理 Heat	未经热处理 None	热处理 Heat	未经热处理 None	热处理 Heat
2019年	26,000	13,000	59,000	15,000	110,000	—
2020年	40,000	15,000	91,000	18,000	170,000	—
2021年	50,000	18,000	114,000	30,000	210,000	—
2022年	62,000	23,000	140,000	40,000	270,000	—
2023年	80,000	30,000	170,000	50,000	320,000	—
2024年	95,000	36,000	200,000	60000	380000	—

3.蓝宝石投资案例
Sapphire Investment Case

　　2021年，我给一位上市公司的客户推荐了一颗13克拉的高品质无烧蓝宝石，当时的价格是90万元人民币。2023年9月，相同品质的蓝宝石在宝石矿区的价格已经达到了145万元人民币。或许是因为特殊时期，或许是因为货币通胀等经济原因导致的涨价。2024年3月我在香港国际珠宝展，全球的珠宝商都会来参展，是一个价格比较公道的针对行业内的展会，都很难看到10克拉的高品质蓝宝石，偶尔看到一颗，价格已经要180万元人民币。

　　In 2021, I recommended a 13-carat high-quality non-burned sapphire to a client of a listed company, and the price at that time was 900,000 yuan. In September 2023, the price of the same quality sapphire in the gem mining area has reached 1.45 million yuan. It may be because of special times, or it may be because of economic reasons such as currency inflation. In March 2024, I was at the Hong Kong International Jewelry Fair, jewelers from all over the world will come to participate in the exhibition with relatively fair price in the industry. It is difficult to see a 10 carat high-quality sapphire. Occasionally I saw one, and the price has been 1.8 million yuan.

Sapphire Investment Case

4. 蓝宝石拍卖纪录
Sapphire Auction Record

时间 Time	拍卖行 Auction House	拍品 Item	成交价 Closing Price
2023年夏拍 Summer, 2023	宝龙伯得富（邦瀚斯） Bonhams	8.17克拉枕形克什米尔蓝宝石的铂金戒指，拥有罕见出色的切割比例，未经过加热处理。 8.17 carat pincushion Kashmiri sapphire platinum ring, with a rare and excellent cut ratio, without heat treatment	估价 35万—55万美元 Estimated USD 350,000-USD550,000
2022年 In 2022	保利香港 Poly Hong Kong	卡地亚设计，重10.45克拉的枕形蓝宝石，产自克什米尔，未经加热。 Cartier design, 10.45 carat pillow-shaped sapphire, made in Kashmir, unheated	成交价 1078.1万港元 Transaction Price HKD 10.781 million
2019年秋拍 Autumn, 2019	佳士得香港 Christie's Hong Kong	蓝宝石戒指，蓝宝石重约16.46克拉，克什米尔产，无加热优化。 Sapphire Ring weighing about 16.46 carats, made in Kashmir, no heating optimization	估价 130万—180万美元 Estimate USD 1.3 million - USD 1.8 million
2018年春拍 Spring, 2018	苏富比纽约 Sotheby's New York	蓝宝石戒指，21.61克拉，产地斯里兰卡，未加热。 Gem Ring, 21.61 carats, Sri Lanka, unheated	成交价 98.7万港元 Transaction Price HKD 987,000

5. 蓝宝石投资收藏建议
Sapphire Investment Collection Suggestions

**小雨老师
经验分享**

Experience Sharing
from Ms. Xiaoyu

（1）权威证书
Authority Certificate

　　选择名贵珠宝投资顾问或者专业度高、可信度高的珠宝商。最好是选择经过权威宝石学院古柏林、SSEF或GRS等国际权威机构认证的蓝宝石。这些证书可以提供关于宝石的详细信息，包括其品质、重量和特征。

　　Choose a precious jewelry investment adviser or a professional and trustworthy jeweler. It is best to choose a sapphire certified by an international authority such as the authoritative gemological institute Gubelin, SSEF or GRS. These certificates can provide detailed information about the gemstone, including its quality, weight and characteristics.

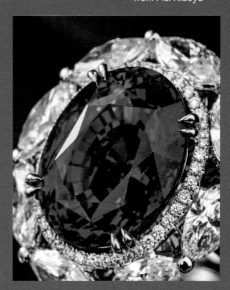

（2）高品质蓝宝石
High-quality Sapphire

　　高品质的蓝宝石往往有更高的保值潜力。关注宝石的颜色、透明度、纯净度和切工等方面，深蓝色、高透明度和少有内部包裹体的宝石往往更受欢迎。

　　High-quality sapphires tend to have a higher potential to preserve their value. Focusing on the color, transparency, purity and cut of the stone, dark blue stones, high transparency and few internal inclusions are often more popular.

（3）市场趋势
Market Trends

　　关注蓝宝石市场的价格和趋势，了解供需关系和买家的偏好。需要注意的是，宝石市场存在波动性，价格受到供求关系、市场趋势和流行趋势等多种因素的影响。因此，在收藏蓝宝石或其他宝石时，建议基于长期投资和兴趣，而非短期炒作和追求快速涨幅。最好咨询专业的名贵珠宝投资顾问。

　　Keep an eye on prices and trends in the sapphire market to understand supply and demand and buyer preferences. It is important to note that there is volatility in the gemstone market, and prices are affected by a variety of factors such as supply and demand, market trends and fashion trends. Therefore, when collecting sapphires or other gemstones, it is recommended to do so based on long-term investment and interest, rather than short-term hype and the pursuit of quick gains. It is best to consult a professional jewelry investment adviser.

　　蓝宝石的投资其实主要还是要从自己的预算来看，比如说你只想投50万元人民币的一颗蓝宝石，我觉得如果想要好出手的话，斯里兰卡的无烧5克拉就很好，如果是有烧的话，可以买一个10克拉的，但如果你有200万元人民币的预算，那我个人建议你买10克拉的无烧的。从保值和投资的角度，50万元跟200万元最大的区别是什么呢？50万元可以变现快一些，200万元的变现慢，但是200万元的放三年五年十年，它的增值的空间要大很多。我一直强调，做宝石投资一定要做时间的朋友，最好找专业的名贵珠宝投资顾问全面地从你的经济情况和宝石价格市场角度帮你定制一个投资策略。

　　Sapphire investment in fact, mainly starts from your own budget. For example, you only want to invest 500,000 yuan of a sapphire, I think if you want to sell better, Sri Lanka's no-burn 5 carat is very good. If it is a burned, you can buy a 10 carat. But if you have a budget of 2 million yuan, then I personally suggest you buy the 10-carat no-burn one. From the perspective of preservation and investment, what is the biggest difference between 500,000 yuan and 2 million yuan? 500,000 yuan can be realized faster, and the realization of 2 million yuan is slower, but in three years, five years and ten years, the appreciation space of 2 million yuan is much larger. I have always emphasized that gem investment must be a friend of time, and it is best to find a professional precious jewelry investment adviser to help you customize an investment strategy from the perspective of your economic situation and gem price market.

Ruby Journey

第二章
02 Chapter Two

红宝石
Ruby

一、红宝石的基本特征
Basic Characteristics of Ruby

红宝石的英文名称为Ruby，是一种珍贵的宝石，是国际流通的五大宝石之一，在国际市场上非常好流通，也是拍卖场上非常受欢迎的名贵珠宝之一，很多国际知名珠宝大品牌都会用到红宝石。主要由氧化铝（Al_2O_3）组成，属于刚玉矿物的一种。微量的铬元素赋予了红宝石典型的红色。红宝石的硬度非常高，莫氏硬度为9，仅次于钻石。这种宝石通常具有良好的光泽和透明度，但也可能包含一些天然的内含物或裂纹。红宝石在不同光源下可能展示不同的红色调，且可能有荧光现象。它是传统上的七月诞生石，长期以来被认为是贵族和皇室的象征。

Ruby, is a precious gemstone, one of the five largest gems in international circulation. It had very good circulation in the international market, and also one of the most popular precious jewelry in the auction house. Many internationally renowned jewelry brands are using ruby. Ruby is mainly composed of alumina (Al_2O_3), which belongs to a corundum mineral. Trace amounts of chromium give rubies their characteristic red color. Rubies are very hard, with a Mohs hardness of 9, second only to diamonds. This gemstone usually has a good luster and clarity, but may also contain some natural inclusions or cracks. Rubies may show different red tones under different light sources, and may have fluorescence. It is the traditional July birthstone and has long been considered a symbol of nobility and royalty.

Basic Characteristics of Ruby

小雨老师宝石交易

小雨老师宝石交易

小雨老师宝石交易

小雨老师宝石交易

小雨老师在矿区

红宝石戒指

红宝石胸针

Basic Characteristics of Ruby

The color of the ruby accounts for 50% of the value of the whole ruby, so the color of the ruby is extremely important.

Ruby

二、如何鉴赏和投资收藏红宝石
How to Appreciate and Invest in Ruby Collection

就红宝石而言，如果要保值增值，颜色必须达到鸽血红，除非它有其他特点突出的点，比如净度格外好，因为红宝石一般都有比较多的裂隙，能达到净度特别好的，是十分难得的。我们鉴赏一颗红宝石主要从7个维度去评价它，只有7个维度的指标都达到好的状态才是一颗精品好宝石。

As far as rubies are concerned, if you want to preserve and increase value, the color must reach pigeon blood, unless it has other characteristics outstanding points, such as its clarity is particularly good. That is because rubies are generally having more cracks, if one can achieve particularly good clarity, it is very rare. We usually evaluate a ruby mainly from 7 aspects. Only if all indicators of the 7 dimensions are achieving good status, it will be considered a high-quality gem.

(一)红宝石的鉴赏
Appreciation of Rubies

1.颜色
Color

是指红宝石有哪些颜色。
It refers to the Color of the Ruby.

红宝石的颜色占据了整颗红宝石价值比例的50%，所以红宝石的颜色极其重要，对市场价格有显著影响。鉴赏红宝石颜色时，主要考虑色调、饱和度和颜色均匀性。最受欢迎的是鸽血红色调，这种色调深且饱和。红宝石的价格随着颜色的深度和饱和度增加而上升，特别是没有任何过渡色调的纯净深红色宝石更为珍贵。颜色均匀、没有色区的红宝石也更受青睐。总的来说，颜色的鲜艳、纯净和均匀性是决定红宝石价值的关键因素。

The color of the ruby accounts for 50% of the value of the entire ruby, so the color of the ruby is extremely important and has a significant impact on the market price. When appreciating ruby color, the main consideration is hue, saturation and color uniformity. The most popular hue is pigeon blood, which is deep and saturated. The price of rubies increases with the depth and saturation of the color, especially pure deep red stones without any transitional tone are more precious. Rubies with uniform color and no color zone are also preferred. In general, the brightness, purity and uniformity of color are the key factors in determining the value of rubies.

颜色不均匀的红宝石

颜色均匀的红宝石

缅甸红宝石
Burmese Ruby

缅甸红宝石以荧光强、鲜艳和深邃的红色而闻名，通常被称为"鸽血红"（Pigeon's Blood）。这种独特的色调是由于较高的铬含量导致的深红色，通常伴有轻微的紫色调。缅甸红宝石的色彩通常非常饱和与均匀，是红宝石中最为珍贵的品种。它的这种特殊色彩使得它在市场上极具价值和受欢迎。

Burmese rubies are known for their intense fluorescence, vivid and deep red color, which is often referred to as "Pigeon's Blood". This distinctive hue is a deep red due to a higher chromium content, and often accompanied by a slight purple tinge. The color of Burmese rubies is usually very saturated and uniform, and it is the most precious variety of rubies. This special color makes it extremely valuable and popular in the market.

缅甸红宝石

Mozambican Ruby

红宝石的鉴赏
Appreciation of Rubies

莫桑比克红宝石

莫桑比克红宝石通常以其鲜艳和明亮的红色著称。这些红宝石的色彩范围从淡红色到深红色不等，有时带有紫色或粉色调。莫桑比克的红宝石因其高度的饱和度和良好的透明度而备受欢迎。虽然它们的色彩可能不如传统的缅甸红宝石那样深，但莫桑比克红宝石以其明亮的色彩和较高的可获得性而在市场上占有一席之地。

Mozambican rubies are usually known for their vivid and bright red color. These rubies' color range from light red to deep red, and sometimes with purple or pink tones. Rubies from Mozambique are popular for their high degree of saturation and good transparency. While they may not be as dark in color as traditional Burmese rubies, Mozambican rubies have a niche in the market for their bright colors and high availability.

莫桑比克红宝石

Appreciation of Ruby

小雨老师经验分享
Experience Sharing from Ms. Xiaoyu

红宝石的颜色占据了整颗红宝石价值的50%，一颗鸽血红色的红宝石比一颗普通红色的红宝石价格可能要相差5倍。在其他条件都相同的情况下，一颗高品质的鸽血红的红宝石，如果是5克拉，它的价格在300万元人民币左右，而一颗普通中浅红色的红宝石，它的价格可能只需要50万元人民币。

The color of the ruby accounts for 50% of the value of the entire ruby, and the price of a dove's blood red ruby may be five times different than that of an ordinary red ruby. On even ground, a high-quality pigeon red ruby, if it is 5 carats, its price is about 3 million yuan, and an ordinary light red, its price may only need 500,000 yuan.

2.净度
Clarity

是指红宝石的内外的纯净程度，是否含有包裹体。
This refers to the purity of the ruby inside and outside, whether it contains inclusions.

1　包裹体的大小、数量和位置 / Size, Quantity and Location of Inclusions

包裹体越大、数量越多，对宝石的净度影响就越大。特别是当包裹体位于台面部位时，对宝石净度的影响更为显著。

The larger and more inclusions, the greater the impact on the clarity of the gem. Especially when the inclusion is located in the mesa, the influence on the clarity of the gem is more significant.

2　包裹体的类型和对比度 / Type and Contrast of Inclusions

包裹体的颜色与红宝石本身的颜色很相近的话，对宝石的净度影响不大。如果包裹体的折射率和颜色与红宝石相差较大，对宝石的净度影响会更明显。

If the color of the inclusion is very similar to the color of the ruby itself, it has little effect on the clarity of the gem. If the refractive index and color of the inclusion are significantly different from the ruby, the impact on the clarity of the gem will be more obvious.

3　未愈合的裂缝 / An Open Fissure

如果宝石内部有未愈合的裂隙，将降低宝石的耐久性，容易受到损伤。

If there are unhealed cracks inside the gem, the durability of the gem will be reduced and it will be vulnerable to damage.

4　丝状包裹体 / Filamentous Inclusions

少量丝状包裹体存在于宝石内部不会影响宝石的净度质量，反而可以改善宝石的外观。

The presence of a small amount of filamentous inclusions inside the gem does not affect the clarity quality of the gem, but can improve the appearance of the gem.

在选择红宝石时，可以尽量选择表面没有矿坑、明显色带、裂隙和深色包裹体的宝石。但是大多数都会有金红石针，只要不影响整体的美观，我们还是可以接受的。原则上，只要肉眼难以看到红宝石内部特征，即可认为宝石具有良好的净度。并不需要追求完全干净，因为完全干净的红宝石反而需要引起怀疑是否为合成的。

In the selection of rubies, you can try to choose gems without pits, obvious color bands, cracks and dark inclusions on the surface. But most of them will have rutile needles, as long as it does not affect the overall beauty, we can still accept. In principle, as long as it is difficult to see the internal characteristics of the ruby with the naked eye, the gem can be considered to have good clarity. It is not necessary to seek perfect cleanliness, because a perfectly clean ruby needs to arouse suspicion of synthetic.

小雨老师经验分享
Experience Sharing from Ms. Xiaoyu

99%的红宝石净度都不可能特别好，内部会有裂隙或者外部有矿缺，如果在社交距离不影响它的整体美观的话，都是可以接受的。即便在佳士得、苏富比等大型国际拍卖的红宝石，净度也不可能非常干净，但是价格依然非常贵。红宝石的净度主要是看它的内部是否有裂隙，以及各种包裹体比如黄铁矿、金红石针。外部主要是看红宝石是否有矿缺，矿缺是否明显影响想到整个宝石的美观，是不是有裂到外面那种对外的裂隙，我们叫做开放性的裂隙。

其实在欧美市场，净度对价格的影响并没有很大，但是在中国市场对净度的要求比较高，高得有点离谱。你会看到在一些拍卖场上的宝石净度并没有那么好，但是依然可以拍出很好的价格。尤其红宝石的净度本身就不好，所以不能要求红宝石的净度跟钻石一样干净，社交距离不影响宝石的美感即可。红宝石天然的裂隙发育就比较明显，所以说干净玻璃体的红宝石是特别少的，打个比方，一颗3克拉的干净玻璃体的红宝石，价格大概在50万元人民币左右，但如果它有点点包裹体，它的价格就会变成45万元人民币，如果包裹体再大一点点，价格可能就变成40万元人民币，当然如果包裹体都影响整体美观了，那就没有价值可言了。净度相对于颜色，净度远远不及颜色对价格的影响那么大。

99% of the clarity of rubies can not be particularly good, there will be cracks inside or outside of the mine. if viewed in the social distance and does not affect its overall beauty, it is acceptable. Even in the large international auction of rubies such as Christie's and Sotheby's, the clarity can not be very clean, but the price is still very expensive. The clarity of rubies mainly depends on whether there are cracks in its interior, and various inclusions such as pyrite and rutile needles. The outside is mainly to see whether the ruby has a lack of ore, whether the lack of ore significantly affects the beauty of the whole gem, whether there is a crack to the outside of the kind of external cracks, and that's what we call it 'open cracks'.

In fact, in the European and American markets, the impact of clarity on the price is not very large, but in the Chinese market, the requirements for clarity are relatively high, which is a little too high. You will see that in some auctions the clarity of the gems is not so good, but it can still fetch a good price. In particular, the clarity of ruby itself is not good, so it is not required that the clarity of ruby is as clean as diamond, as long as viewed in social distance there is no affect on the beauty of the gem. The natural crack development of rubies is more obvious, so the clean vitreous ruby is particularly few. For example, a 3-carat clean vitreous ruby, the price is about 500,000 yuan, but if it has a little inclusion, its price will become 450,000 yuan, and if the inclusion is a little larger, the price may become 400,000 yuan. Of course, if the inclusions affect the overall beauty, there is no value at all. Relative to color, clarity is far less impact than color on price.

3.切工
Cut

是指红宝石外表切割的对称度和工整度。
It refers to the symmetry and smoothness of the ruby cutting.

切工主要表现在对称性和抛光质量等方面。对称性的问题指正侧面轮廓的对称性偏差、台面偏心、底尖偏心、亭部膨胀、刻面畸形和刻面尖点不尖等。抛光质量指的是宝石表面的抛光程度。红宝石的切割对其价格有显著影响。高质量的切割不仅可以增强红宝石的颜色和光泽,还可以最大化其火彩和闪光效果,从而提升整体美感和价值。切割的质量影响宝石对光的反射和折射,进而影响其光泽和火彩,也就是说切割影响到红宝石的闪耀程度。专业的切割技术能够精准地展现宝石的最佳面貌,包括颜色分布和内部包裹体的展示。精心切割的红宝石通常会比粗糙或非标准切割的同质量红宝石售价更高。然而,过度的切割可能导致宝石重量的减少,这也可能影响其市场价值。总的来说,切割是红宝石价值的一个重要组成部分。

The cutting is mainly manifested in symmetry and polishing quality. The symmetry problems include the symmetry deviation of the side profile, the eccentricity of the mesa, the eccentricity of the bottom tip, the bulge of the pavilion, the facet deformity and the non-cusp of the facet. Polishing quality refers to the degree of polishing of the gem's surface. The cutting of rubies has a significant effect on their price. High-quality cutting not only enhances the color and shine of the ruby, but also maximizes its fire and sparkle effect, thereby enhancing the overall beauty and value. The quality of the cut affects the reflection and refraction of the stone to the light, which in turn affects its luster and fire color, that is to say, the cutting affects the shine of the ruby. Professional cutting techniques accurately show the best of the stone, including the distribution of color and the display of internal inclusions. Carefully cut rubies will usually command a higher price than rough or non-standard cuts of the same quality. However, excessive cutting can result in a reduction in the weight of the stone, which can also affect its market value. Overall, cutting is an important component of a ruby's value.

小雨老师经验分享
Experience Sharing from Ms. Xiaoyu

红宝石露底或者切割太厚都会直接影响价格。露底指红宝石中心颜色很淡,一般是由台面过大引起的。红宝石切得太厚,就会让台面变小,而且会有黑域出现,影响到闪烁程度,从而影响红宝石的价格。最好的红宝石的切割就是台面舒展、火彩反光好。红宝石的价格跟形状没有太大关系,圆形的红宝石不一定是最贵的,跟钻石完全不一样,钻石圆形会更贵一些。红宝石各种形状都有,一般垫型和椭圆形会更讨喜一些,价格也会好一些。如果一颗3克拉的红宝石,但是台面小,看着像2克拉,你就会觉得不划算,我们可以将宝石重新切割。

Ruby's bare bottom or cut too thick will directly affect the price. Bare bottom refers to the ruby center color is very light, generally caused by too large platform. If the ruby is cut too thick, it will make the table platform, and there will be black areas, which will affect the degree of flicker, and thus affect the price of the ruby. The best ruby cutting is the platform stretch, and reflective good fire color. The price of ruby does not have much to do with the shape, the round ruby is not necessarily the most expensive, and the diamond is completely different, the round diamond will be more expensive. Ruby has a variety of shapes, the general pad type and oval type will be more pleasing, and the price will be better. If you have a 3-carat ruby with a small countertop that looks like 2 carats, you're not going to get a good deal, and we can cut the gem again.

红宝石漏底图

红宝石厚底图

4.克拉重量
Carat Weight

红宝石的克拉重量对其价格有显著影响。一般来说，克拉重量越大，单克拉价格越高。这是因为大克拉重量的红宝石更为稀有，且更能展现出宝石的美丽和特点。当红宝石的重量增加时，其价格通常以指数级增长。例如，1克拉的高品质红宝石可能会非常昂贵，但2克拉的同等品质红宝石的价格可能会比两个1克拉红宝石的总和更高。

The carat weight of a ruby has a significant effect on its price. Generally speaking, the higher the carat weight, the higher the price per carat. This is because large carat-weight rubies are rarer and better able to show the beauty and character of the stone. As the weight of a ruby increases, its price usually increases exponentially. For example, a 1-carat high-quality ruby can be very expensive, but a 2-carat ruby of the same quality can cost more than two 1-carat rubies combined.

小雨老师经验分享
Experience Sharing from Ms. Xiaoyu

红宝石是有克拉溢价的，打个比方，1克拉的红宝石5万元人民币，3克拉的红宝石可能要60万元人民币，6克拉的红宝石可能需要280万元人民币，10克拉的红宝石可能需要2500万元人民币。如果作为红宝石保值增值而言，3克拉的高品质红宝石都值得收入囊中。5克拉以上，更是投资精品，但是市面上5克拉以上的高品质好货很难得。

For example, a 1-carat ruby may cost 50,000 yuan, a 3-carat ruby may cost 600,000 yuan, a 6-carat ruby may cost 2.8 million yuan, and a 10-carat ruby may cost 25 million yuan. If it is used as a ruby preservation and appreciation, 3 carat high-quality rubies are worth earning. If more than 5 carats, it is a boutique investment, but in the market, a ruby that weight more than 5 carats and of high-quality is very rare.

5.产地
Place of Origin

是指红宝石的出产地，对价格影响挺大的。
It refers to the origin of the ruby, which has a big impact on the price.

莫桑比克
Mozambique

莫桑比克红宝石现在整体价格还不高，还是投资的洼地。莫桑比克现在有很多高品质的红宝石，整体效果要比缅甸的红宝石刚性一些，净度也要好很多，目前非常受市场欢迎，整体价格没有缅甸高，很值得投资。

"Estrela de Fura"（意为"Fura之星"）红宝石：它在2023年6月8日的纽约苏富比拍卖会上以3,480万美元的价格售出，创下了单颗红宝石拍卖总价的世界纪录。这颗红宝石原始毛坯重达101克拉，经过打磨和抛光后，成为一颗55.22克拉的宝石，其珍贵程度和显著的光学透明度赢得了"一生一次"的珠宝称号。这次拍卖标志着莫桑比克红宝石在国际舞台上的崛起，并证明了其最优质的红宝石可以与世界上任何产地的红宝石相媲美。这颗红宝石的发现地是莫桑比克北部的Montepuez地区，这里是世界上最富产的红宝石矿区之一。这个地区的红宝石被誉为具有"鸽血红"色调，这是一种通常只与缅甸红宝石相关联的珍贵颜色。由于其出色的清晰度和鲜艳的红色，这颗红宝石被认为是非洲宝石中的佼佼者，能够与缅甸红宝石竞

Mozambique ruby's overall price is not high yet, which is an investment hollow land. Mozambique now has a lot of high-quality rubies, the overall appearance is more rigid than Myanmar's ruby, and the clarity is also much better. Thus it is currently very popular in the market, and the overall price is not high in Myanmar, so it is worth investing.

"Estrela de Fura" (meaning "Star of Fura") ruby: It sold for $34.8 million at Sotheby's New York on June 8, 2023, setting a world record for the total price paid at auction for a single ruby. This ruby, which weighs 101 carats in its original rough, has been polished to become a 55.22-carat gem, earning the title of "once in a lifetime" jewel for its precious and remarkable optical transparency. This auction marks the rise of Mozambican rubies on the international stage and proves that its finest rubies can rival those of any origin in the world. The ruby was found in the Montepuez region of northern Mozambique, one of the richest ruby mining areas in the world. Rubies from this region are reputed to have a "pigeon blood" hue, a precious color usually associated only with Burmese rubies. Due to its outstanding clarity and vivid red color, this ruby is considered a standout among African gemstones, which is able to compete with the Burmese ruby.

坦桑尼亚Winza
Tanzania Winza

Winza红宝石产自坦桑尼亚的Winza地区，是一种非常特别且稀有的宝石。它具有一些独有的特征，让它在宝石界中占有一席之地，尽管其开采时间相当短暂。小雨老师特别提醒：Winza红宝石这个矿区的红宝石真的非常漂亮，晶体非常好，但是太稀少了，大克拉的很有收藏价值。

Winza红宝石的价格信息：因其出色的品质和独特性在市场上非常受欢迎。高品质的Winza红宝石，尤其是未经加热处理的，每克拉售价可高达25,000美元。然而，大多数Winza红宝石的价格在每克拉4,000至8,000美元之间。这些红宝石的显著特征包括蓝色色带，以及它们独特的透明度和光彩。Winza红宝石的开采和市场表现表明，即使是在红宝石领域中，它们也是一种珍稀和值得收藏的宝石。

The Winza ruby, from the Winza region of Tanzania, is a very special and rare gemstone. It has some unique characteristics that have given it a place in the gemstone world, despite its very short mining time. Teacher Xiaoyu kindly remind you that: The ruby of Winza mining area is really very beautiful, the crystal is very good, but it is too rare, and the large carat is very valuable for collection.

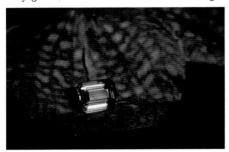

Winza红宝石

Price information of Winza rubies: They are very popular in the market because of their excellent quality and uniqueness. High-quality Winza rubies, especially unheated ones, can sell for as much as $25,000 per carat. However, most Winza rubies are priced between $4,000 and $8,000 per carat. Notable features of these rubies include blue bands, as well as their unique transparency and brilliance. The mining and market performance of Winza rubies show that they are a rare and collectible gem, even in the field of rubies.

缅甸 Myanmar

缅甸是世界上非常著名的宝石产地，特别以其红宝石闻名。鸽血红这种颜色就来自缅甸产区，红色很像鸽子的眼睛。高品质的缅甸红宝石颜色是那种浓浓的像血一样的绒质感，带有一点点玫色调，有强荧光。但是也有完全玫红色的缅甸红宝石，通常晶体不够好，价格也比较低廉。缅甸的地质构造非常适合红宝石的形成。这些宝石主要在大理石中形成，这种岩石为红宝石的颜色和质地提供了理想的条件。缅甸的红宝石因其深沉且饱和的红色、均匀的纹理和丰富的色彩层次而备受推崇。缅甸的主要红宝石矿区包括抹谷（Mogok）和孟素（Mong Hsu）。缅甸的红宝石在国际市场上非常受欢迎，经常出现在拍卖场所，并被认为是目前红宝石中最贵的品种之一。例如，一枚8.99克拉的红宝石和钻石戒指在2013年的一次拍卖中以3,935,105美元的价格售出，相当于每克拉437,720美元。这枚红宝石来自历史悠久的抹谷矿区，并以其稀有的"鸽血"色调而闻名。

Myanmar is a very famous gemstone producer in the world, especially for its rubies. Pigeon blood is a color that comes from Burma, and the red color resembles the eye of a pigeon. The high quality Burmese ruby color is a thick, blood-like velvet texture with a slight rose tone and strong fluorescence.

红宝石戒指

But there are also Burmese rubies that are completely rose red, which are usually not good enough crystal, and the price is lower. Burma's geology is perfect for ruby formation. These gems are formed primarily in marble, a rock that provides ideal conditions for the color and texture of rubies. Myanmar rubies are highly propmoted for their deep and saturated red color, uniform texture and rich color layers. Major ruby mining areas in Myanmar include Mogok and Mong Hsu. Rubies from Myanmar are very popular in the international market, and often appearing at auction venues, and are considered to be one of the most expensive varieties of rubies currently available.

For example, an 8.99-carat ruby and diamond ring sold for $3,935,105 at an auction in 2013, which equates to $437,720 per carat. This ruby comes from the historic Mogok mining area and is known for its rare "pigeon blood" hue.

Sum up 总结

红宝石的产地对价格有显著影响。缅甸和莫桑比克产的红宝石通常更为珍贵，因其颜色、透明度和历史背景。其他产地如泰国、斯里兰卡等的红宝石价格相对较低。产地影响了宝石的品质、稀缺性，从而影响了市场价值。大家可以参考佳士得、苏富比、保利等国际珠宝拍卖公司，他们主要拍卖的都是缅甸红宝石和莫桑比克红宝石，尤其是缅甸天然的无烧红宝石，10克拉的价格可以拍到2500万元左右。

The origin of a ruby has a significant effect on the price. Rubies from Myanmar and Mozambique are often more valuable because of their color, clarity and historical background. Other sources such as Thailand, Sri Lanka and other ruby prices are relatively low. The origin of the stone affects the quality, scarcity and thus the market value. We can refer to Christie's, Sotheby's, Poly and other international jewelry auction companies. Their main auction is Myanmar ruby and Mozambique ruby, especially Myanmar's natural non-burned ruby, where 10 carats price can be auctioned about 25 million yuan.

小雨老师经验分享
Experience Sharing from Ms. Xiaoyu

产地的声誉和稀缺性是影响红宝石价格的关键因素。例如，一颗来自缅甸的红宝石在同等品质情况下，价格通常要比来自莫桑比克的红宝石高出5倍，甚至更多。以6克拉的红宝石为例，来自缅甸的高品质红宝石的价格大约在一千八百万元人民币左右，而来自莫桑比克的红宝石价格大约在四百万元人民币左右。前10年，缅甸的红宝石在拍卖场上有绝对的优势，按照单克拉计算，每克拉缅甸红宝石价格有的可以超过一百万美元，而莫桑比克红宝石单克拉在二十万美元左右。所以莫桑比克红宝石和缅甸红宝石至少有5倍的差距。但是这两年开始，莫桑比克的红宝石价格涨得非常明显。2023年，在美国苏富比拍卖的一颗55.22克拉的莫桑比克红宝石，最后以3480万美元的价格成交。这不仅打破了红宝石的拍卖纪录，也打破了莫桑比克红宝石的拍卖纪录。

福荣之星

The reputation and scarcity of the origin are key factors affecting the price of rubies. For example, a ruby from Myanmar is usually five times or more expensive than a ruby from Mozambique of the same quality.

红宝石戒指

Taking 6-carat rubies as an example, the price of high-quality rubies from Myanmar is about 18 million yuan, and the price of rubies from Mozambique is about 4 million yuan.

In the first 10 years, Myanmar's ruby has an absolute advantage in the auction market, and the price of Myanmar ruby per carat can exceed one million US dollars, while the Mozambique ruby single carat is about 200,000 US dollars. So there's at least a five times difference between the Mozambican ruby and the Burmese ruby. But in the past two years, the price of rubies in Mozambique has risen significantly. In 2023, a 55.22-carat Mozambican ruby auctioned at Sotheby's in the United States was finally sold for $34.8 million. This not only broke the auction record for rubies, but also broke the auction record for rubies from Mozambique.

RUBY

6.优化
Treatment

Ruby

是指这颗红宝石是否天然，是否经过优化处理。
It means whether the ruby is natural, whether it has been treated.

热处理
Heat Treatment

在宝石鉴定学中，对于红宝石的优化是国际上认可的，但是红宝石的处理是国际上不认可的。以下我们指出的红宝石热优化（"老烧"），是国际上认可的，日常可消费的。而红宝石的染色处理，填充处理是国际上不认可的，如果卖给消费者要明确说明，这颗红宝石是处理的。作为红宝石投资收藏而言，自己预算充足的情况下，我们尽量买大克拉天然无烧的，如果作为保值增值而言，买大克拉有烧的也可以。不能买处理的红宝石，不管多大克拉，都没有市场价值。红宝石的染色处理和红宝石的填充处理都是国际上不认可的，证书上也会标明。红宝石是否经过热处理（"有烧"或"无烧"）对其价格有显著影响。

红宝石的热优化技术"老烧"是一种常用的宝石优化方法，目的是改善或增强红宝石的颜色和透明度。在这个过程中，红宝石被加热到高温，通常在几百到几千摄氏度之间，持续一段时间。热处理可以减少或消除红宝石内部的某些瑕疵和不纯物质，从而使颜色更加鲜艳和均匀。这种处理是永久性的，被认为是宝石行业的标准做法。在消费者购买时，诚实标明红宝石是否经过热处理是非常重要的，因为这直接影响着宝石的价值和消费者的购买决策。对于那些寻求天然未经处理宝石的消费者来说，了解热处理的存在和其对宝石特性的影响尤为重要。尤其是作为投资，最好选择天然无烧的鸽血红宝石。

In gemology, the optimization of rubies is internationally recognized, but the treatment of rubies is not internationally recognized. Below we point out the ruby heat optimization (" old burn "), which is internationally recognized and daily consumable. The dyeing treatment and filling treatment of rubies are not recognized internationally. If it is sold to consumers, it should be clearly stated that this ruby is treated. As a ruby investment collection, in the case of sufficient budge, we try to buy large carat natural without burning, and if as a value preservation, buying large carat burned can also be good. You should not buy processed rubies, no matter how big the carats are, they have no market value. The dyeing treatment of rubies and the filling treatment of rubies are not recognized internationally, and will be indicated on the certificate. Whether a ruby has been heat treated (" burnt "or" unburnt ") has a significant impact on its price.

"Old burning" is a commonly used gem optimization method to improve or enhance the color and transparency of rubies. In this process, rubies are heated to high temperatures, usually between hundreds and thousands of degrees Celsius, for a period of time. Heat treatment can reduce or eliminate certain defects and impure substances inside the ruby, thus making the color more bright and even. This treatment is permanent and is considered standard practice in the gemstone industry. At the time of purchase, it is very important to be honest about whether the ruby has been heat treated, as this directly affects the value of the gem and the purchasing decision of the consumer. For those consumers seeking natural, untreated gemstones, it is particularly important to understand the existence of heat treatment and its impact on gemstone properties. Especially as an investment, it is best to choose natural non-burned pigeon blood gems.

未热处理
Unheat Treated

未经热处理（"无烧"）的红宝石通常比经过热处理的红宝石（"有烧"）更为珍贵和昂贵，因为它们更加稀有且被视为更纯净和自然。热处理可以改善红宝石的颜色和透明度，但也会降低其稀缺性和收藏价值。因此，未经热处理的红宝石通常在市场上更受欢迎，尤其是在珠宝收藏家和高端市场中。例如，一些高品质的未经热处理的红宝石在国际拍卖会上可以以每克拉数十万美元的价格成交，甚至更高。这反映了市场对纯净自然宝石的高度评价和需求。（备注：红宝石的"新烧"也就是工业烧，是国际上不认可的。）

Rubies that have not been heat-treated (" unburned ") are generally more precious and expensive than rubies that have been heat-treated (" burned ") because they are rarer and seen as more pure and natural. Heat treatment can improve the color and transparency of rubies, but it can also reduce their scarcity and collectible value. Therefore, rubies that have not been heat treated are generally more popular in the market, especially among jewelry collectors and the high-end market. For example, some high-quality, rubies with unheated treatment can sell at international auctions for hundreds of thousands of dollars per carat, or even more. This reflects the high value and demand for pure natural gemstones in the market. (Note: The "new burning" of rubies, that is, industrial burning, is not recognized internationally.)

热处理红宝石

莫桑比克红宝石
从左到右：未处理、加热、加热和填充铅玻璃的裂缝

小雨老师经验分享
Experience Sharing from Ms. Xiaoyu

其实在欧美市场大多数是有烧的红宝石，因为佩戴更有性价比，他们对于红宝石的有烧无烧跟国内的市场接受度是很不相同的。在欧美国家，大多数人都能够接受红宝石的有烧，而且大克拉的有烧的红宝石一点儿都不便宜。像无烧的红宝石，在欧美市场还不如有烧的红宝石销售得好。因为一颗无烧的红宝石可能要5万元人民币，但是一颗有烧的红宝石只需要一两万元人民币，都是小克拉的话，其实有烧无烧都无所谓，因为只是一个佩戴作用。但是从保值来讲的话，在欧美市场，大克拉的有烧的特别漂亮，那么它也是有保值增值的潜力。但如果是收藏投资的话，那一定是无烧的大克拉，因为更稀缺，价值会更高一些。

In fact, most of the markets in Europe and the United States, there are burned rubies, because they are more cost-effective to wear, and they are very different from the domestic market in terms of acceptance of rubies. In Europe and the United States, most people can accept the burning of rubies, and large carat burned rubies are not cheap at all. Like unburned rubies, they sell worse in the European and American markets than burned rubies. Because a ruby without burning may cost 50,000 yuan, but a ruby with burning only needs 20,000 yuan, if it is a small carat, in fact, it does not matter whether it is burned or not, because it is only a wearing effect. However, in terms of value preservation, in the European and American markets, the large carat is particularly beautiful, so it also has the potential to maintain and increase value. But if it is a collection investment, it must be unburned large carats, because more scarce, the value will be higher.scarce value will be higher.

7.证书
Certificate

不同的证书代表不同的含金量。
Different certificates represent different gold content.

01
Gubelin

古柏林实验室是一家在宝石鉴定和研究领域享有高度声誉的瑞士实验室。在拍卖场上经常出现，含金量非常高。如果宝石品质足够好，也会有专门的铂金证书。它尤其以对彩色宝石（如红宝石、蓝宝石和祖母绿）的深入分析而著称。古柏林证书不仅提供宝石的基本信息，如重量、尺寸和颜色，还提供详细的产地分析和是否经过任何形式的处理。对于那些寻求高级和详细宝石证书的收藏家和投资者来说，古柏林实验室的证书是一个可靠的选择。小雨老师特别提醒：古柏林对红宝石颜色评级很少给到鸽血红，但是会对整个宝石打分，比如90分以上就是很不错的宝石。

小雨老师在鉴定宝石

古柏林证书

Gubelin Laboratory is a Swiss laboratory with a high reputation in the field of gemstone identification and research. It is often found at auction, and the gold content is very high. If the gem quality is good enough, there will also be a special platinum certificate. It is particularly notable for its in-depth analysis of colored gemstones such as rubies, sapphires and emeralds. The Gubelin certificate provides not only basic information about the gemstone, such as weight, size and color, but also a detailed analysis of its origin and whether it has been treated in any way. For those collectors and investors seeking certificates in high-grade and detailed gemstones, certificates from Gubelin Laboratories are a reliable choice. Reminder from Ms. Xiaoyu: Gubelin rarely gives ruby color rating for pigeon's blood red, but they will score the entire gem, such as with 90 points or more, it is a very good gem.

036 | Luxury Jewelry

Certificates

02 SSEF

来自瑞士，非常权威的国际证书，在拍卖场上经常出现，含金量非常高。会对宝石的重量、产地、有无加热、颜色做分级。如果产自缅甸、阿富汗、越南、达到鸽血红颜色，晶体也好的，会给到鸽血红颜色。但莫桑比克的最好颜色描述就是red with strong saturation。小雨老师特别提醒：因为SSEF的证书特别贵，如果这颗红宝石很好的话，就会考虑配一张SSEF的证书，但是作为宝石投资者而言，千万不能看见这颗红宝石配了一张SSEF的证书，就给很高的价格。因为有时候，证书只能体现宝石的颜色，并不能说明它的净度，所以有些珠宝商就故意钻这个空当，将一个颜色很好但是净度和切工都很普通的红宝石配一张SSEF的证书，然后忽悠宝石爱好者，卖一个很高的价格。

A very authoritative international certificate from Switzerland, often found at auction, with a very high gold content. The gems are graded for weight, origin, heating, and color. If it is produced in Myanmar, Afghanistan, Vietnam, and reaches the color of pigeon blood, the crystal is also good, and the color of pigeon blood will be given. But the best color description of Mozambique is red with strong saturation. Reminder from Ms. Xiaoyu: As the SSEF certificate is particularly expensive, if the ruby is very good, we will consider matching a SSEF certificate. But as a gem investor, we must not give a high price just seeing a SSEF certificate of red ruby. Because sometimes, the certificate can only reflect the color of the gem, and can not explain its clarity, so some jewelers deliberately drill this gap, match a good color but very ordinary clarity and cut ruby with a SSEF certificate. And then they fool gem lovers to sell a high price.

03 AGL

American Gemological Laboratories

来自美国，是非常严苛的一家宝石实验室。在欧美市场的含金量很高，也会出现在拍卖场上。会对宝石的重量、产地、有无加热、颜色做分级。在鉴定颜色宝石方面非常有声誉，提供详尽的红宝石信息，比如是否有热优化，来自哪个产地，但是它很少对红宝石有颜色评级为鸽血红。这家机构跟瑞士的古柏林很像，会对这颗宝石的整体表现打分，比如90分、95分。其实这样的高分也是对这颗宝石的整体评价，对于这颗宝石的交易也是一种保障。

It's from the United States. It's a very rigorous gemstone lab. The gold content in the European and American markets is very high and will also appear in the auction. The gems are graded for weight, origin, heating, and color. It has a reputation for identifying colored gems, providing detailed information about rubies such as whether they are thermally optimized and where they come from, but it rarely gives color ratings to rubies as pigeon blood. The agency, much like Gubelin in Switzerland, gives the gem a score of 90 or 95 for its overall performance. In fact, such a high score is also the overall evaluation of this gem, and it is also a guarantee for the trading of this gem.

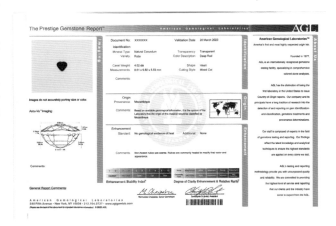

AGL红宝石证书

SSEF鸽血红证书

04
GRS

来自瑞士的宝石实验室，主要做彩色宝石的鉴定，含金量也很高，也会出现在拍卖场上，如果宝石品质足够好，还会有铂金证书。鉴定机构对于红宝石的颜色，分为副证鸽血红和主证鸽血红，副证和主证的区别就是，主证鸽血红颜色偏甜一点，会带有一点枚粉色调，荧光强一些。鸽血红就是红色达到一定的饱和度，但是如果颜色红得发黑，那么市面上叫黑鸽子血，这样的价格往往比较便宜。真正的好的鸽血红颜色是浓郁而不是发黑。小雨老师特别想提醒一下，千万不要产生一个误区，主证的鸽血红就要比副证的鸽血红贵一些。副证的鸽血红颜色也是非常好的，非常有价值。带主证只能说明这颗颜色偏甜一点，荧光强一些，主要是缅甸出产的红宝石，还有一些莫桑比克和马达加斯加的红宝石也 有可能拿到主证鸽血红。总之，主证不能说明这个颜色是最好最贵的，一颗红宝石的价格取决于几个维度，不能光看一个颜色。

It is a gemstone laboratory from Switzerland, mainly doe the identification of colored gems, the gold content is also high, will also appear in the auction. If the gemstone quality is good enough, there will be a platinum certificate. For the color of ruby, identification agencies divided them into secondary pigeon blood and the main pigeon blood, where the difference between the secondary and the main evidence is that the main pigeon blood color is a little sweeter, having a little pink tone, and the fluorescence is stronger. Pigeon blood is red to a certain degree of saturation, but if the color is red and black, then the market calls it black pigeon blood, so the price is often cheaper. Really good pigeon blood color is rich red rather than black. Ms. Xiaoyu wants to especially remind that, do not having a misunderstanding, that the main certificate of pigeon blood is more expensive than the deputy certificate of pigeon blood. The color of pigeon blood is also very good and valuable. With the master certificate can only show that the color is a little sweeter and the fluorescence is strong. Rubies majorly produced in Myanmar, and some rubies from Mozambique and Madagascar may also get the master certificate pigeon blood. In short, the main certificate can not show that this color is the best and most expensive, the price of a ruby depends on several aspects, we can not just judge by a color.

GRS红宝石证书

05
GIA
Gemological Institute of America

来自美国，主要做钻石的鉴定，也可以做红宝石的鉴定。提供全面的宝石分析，包括是否经过热优化，但是对于颜色这一块很少会给皇家蓝和鸽血红，一般是缅甸的红宝石才会给鸽血红，所以颜色分级这块在GIA的证书上体现得没有那么商业化。

From the United States, mainly does diamonds. It can also do ruby identification. It provides full gemstone analysis, including whether or not the gemstone has been thermally optimized. But in terms of color, it rarely gives royal blue and pigeon blood certificate, which is generally given to Burmese rubies, so the color grading is less commercialized on the GIA certificate.

GIA红宝石证书

06
AIGS
Asian Institute of Gemological Sciences

亚洲宝石学院是一家总部位于泰国曼谷的宝石学院。该学院专注于提供宝石和珠宝的培训、鉴定和研究服务。学院的宝石鉴定和学术研究在亚洲地区享有良好声誉。亚洲宝石学院对于红宝石的颜色会有鸽血红评级。

The Asian Gemological Institute is a gemological institute headquartered in Bangkok, Thailand. The College focuses on providing training, identification and research services in gemstones and jewelry. The school's gemstone identification and academic research enjoys a good reputation in the Asian region. The Asian Gemological Institute gives a pigeon blood rating for the color of rubies.

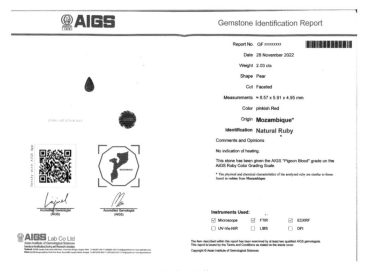

AIGS红宝石证书

小雨老师经验分享
Experience Sharing from Ms. Xiaoyu

所有的证书只是参考，因为证书上是没法显示切工和净度的好坏的，瑞士的古柏林证书和美国的AGL证书会对这颗宝石综合打分，也就是整体评价，可以参考一下，但最终还是看每颗宝石本身，每颗宝石都不同，选择这些知名实验室的证书可以部分确保您所购买的红宝石符合收藏级别的品质和真实性。大家不能光看一个证书去评价这个宝石的好坏，因为有的珠宝商给一颗普通的宝石，去做了一个古柏林或者SSEF的证书，因为那个证书的含金量比较高，很多不懂的消费者就会觉得那证书那么贵，这个宝石一定很好，其实不一定，有可能珠宝商就利用了这一点。所以说我们最终还是看这个宝石本身，而不能用一张证书就评定了它的价格。

All the certificates are only for reference, because the certificate can not show the quality of cut and clarity. The Swiss Gubelin certificate and the United States AGL certificate will be a comprehensive score of this gem, which is the overall evaluation. You can use them as reference, but in the end you have to look at each gem itself, as each gem is different. Choosing a certificate from one of these renowned laboratories partially ensures that the ruby you purchase meets collectible quality and authenticity. We can not just look at a certificate to evaluate the quality of this gem, because some jewelers give an ordinary gem, to do a Gubelin or SSEF certificate, because the gold content of the certificate is relatively high. Many consumers not understand the certificate will feel that the certificate is so expensive, the gem must be very good, but in fact, it is not necessarily right. There may be jewelers taking advantage of this. So in the end, we look at the stone itself, and we can't value it with a certificate.

（二）红宝石的投资收藏建议
Rubies Investment and Collection Suggestions

1.红宝石涨价的原因
Reasons for the Price Rise of Rubies

　　产量低： 2018年我去缅甸考察红宝石开采市场，大多数红宝石裂隙发育都特别明显，大多数都是低劣品质，能达到玻璃体的鸽血红红宝石非常少。有时候到矿区你看到好几颗高品质的红宝石，其实并不是当月开采出来的，可能是一年开采出来的量积累下来的。矿工有时候一个月都开采不到1颗高品质的红宝石。所以1克拉的高品质缅甸红宝石都非常稀有，3克拉、5克拉、10克拉更是稀有，所以缅甸的红宝石价格非常高。还有一个产区叫做莫桑比克，莫桑比克的出产量相比缅甸要高一些，但是出产量也在逐年下降。2克拉以上的高品质莫桑比克红宝石在市场上非常受欢迎，3克拉、5克拉、10克拉的高品质红宝石涨幅幅度非常明显。70%的红宝石原石是通过GEMFIELDS宝石矿业公司拍卖进入市场的，以下是2018年到2021年红宝石原石拍卖的价格。从2018年的5500万美元涨到2021年的8800万美元，红宝石原石的小幅度涨价都会引起红宝石成品的翻倍涨价。

　　Low production: In 2018, I went to Myanmar to investigate the ruby mining market, and most of the ruby cracks were particularly obvious. Most of them were of poor quality, and very few pigeon red rubies could reach the vitreous body. Sometimes in the mining area you see several high-quality rubies, but in fact, it is not mined in this month. It may be the accumulation of the amount mined in a year. Miners sometimes produce less than one high-quality ruby a month. So 1 carat high-quality Myanmar rubies are very rare, and 3 carat, 5 carat, 10 carat are rarer, so the price of Myanmar rubies is very high. There is another producing region called Mozambique, which has a higher production than Myanmar, but the production is also declining year by year. High-quality Mozambican rubies above 2 carats are very popular in the market, and the increase in high-quality rubies of 3 carats, 5 carats and 10 carats is very obvious. 70% of rough rubies come to market through GEMFIELDS auctions, and here are the prices of rough rubies at auctions from 2018 to 2021. From $55 million in 2018 to $88 million in 2021, a small increase in the price of the raw rubies will cause double price increase of the finished.

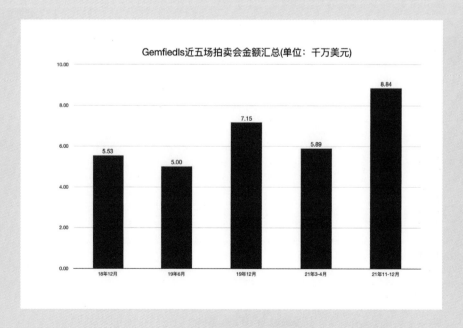

2. 红宝石近几年的价格涨幅
The Price Increase of Rubies in Recent Years

缅甸红宝石近5年价格涨幅
The Price Increase of Myanmar Rubies in the Past 5 Years

克拉 Carat	2.00—3.00克拉 (元人民币/克拉) 2.00—3.00ct (CNY/ct)	5.00克拉 (元人民币/克拉) 5.00ct (CNY/ct)	10.00克拉 (元人民币/克拉) 10.00ct (CNY/ct)	20.00克拉 (元人民币/克拉) 20.00ct (CNY/ct)
处理 Treatment	未热处理 None	未热处理 None	未热处理 None	未热处理 None
2019年	133,000	357,000	892,000	1,338,000
2020年	200,000	550,000	1,370,000	2,069,000
2021年	257,000	690,000	1,716,000	2,570,000
2022年	321,000	858,000	2,145,000	3,210,000
2023年	378,000	1,009,000	2,523,000	3,780,000
2024年	420,000	1,200,000	2,600,000	—

莫桑比克红宝石近5年价格涨幅
Mozambican Ruby Price Increase in the Past 5 Years

克拉 Carat	2.00—3.00克拉 (元人民币/克拉) 2.00—3.00ct (CNY/ct)	5.00克拉 (元人民币/克拉) 5.00ct (CNY/ct)	10.00克拉 (元人民币/克拉) 10.00ct (CNY/ct)	20.00克拉 (元人民币/克拉) 20.00ct (CNY/ct)
处理 Treatment	未热处理 None	未热处理 None	未热处理 None	未热处理 None
2019年	44,000	119,000	297,000	446,000
2020年	68,000	183,000	457,000	686,000
2021年	86,000	220,000	572,000	858,000
2022年	107,000	286,000	715,000	1,072,000
2023年	126,000	356,000	841,000	1,261,800
2024年	150,000	380,000	900,000	—

3. 红宝石拍卖纪录
Ruby Auction Record

时间 Time	拍卖行 Auction House	拍品 Item	成交价 Closing Price
2023年4月 April, 2023	苏富比香港 Sotheby's Hong Kong	8.05克拉的软垫形缅甸红宝石配钻石戒指 未经过加热处理 8.05 carat padded Burmese ruby with a diamond ring, without heat treatment	成交价 14,970,000 港元 Sale Price HKD 14,970,000
2023年4月 April, 2023	苏富比香港 Sotheby's Hong Kong	20.19克拉和18.68克拉，海瑞温斯顿设计缅甸红宝石耳坠 未经加热 20.19 and 18.68 carats, designed by Harry Winston, Burmese ruby earrings, without heat treatment	成交价 15,575,000 港元 Sale Price HKD 15,575,000
2020年10月 October, 2020	佳士得日内瓦 Christie's Geneva	13.01克拉天然缅甸红宝石配钻石戒指 未经过加热处理 13.01 carat natural Burmese ruby with diamond ring sold without heat treatment	成交价 942,000 法郎 Sale Price CHF 942,000
2019年秋拍 The Fall 2019 auction	保利北京 Poly Beijing	5.45克拉椭圆形切割莫桑比克鸽血红红宝石戒指，主石未经加热 5.45 carat oval cut Mozambique pigeon red Ruby, the main stone without heat treatment	成交价 644,000 人民币 Sale Price CNY 644,000

4. 红宝石投资案例
Ruby Investment Case

2018年我在曼谷考察红宝石市场，有一个朋友是一个珠宝商，但是她主要做低端宝石，我就给她建议，最好把赚的钱用来买名贵珠宝，保值能力会更强。我在曼谷给她推荐了一颗5克拉的高品质红宝石，当时价格在40万元人民币。在2023年，她的这颗5克拉的红宝石卖了150万元人民币。

In 2018, I visited the ruby market in Bangkok. A friend of mine was a jeweler, but she mainly did low-end gems, so I gave her advice that it was best to use the money to buy luxury jewelry, whose ability to maintain value would be stronger. I recommended a high-quality 5-carat ruby to her in Bangkok, which was priced at 400,000 yuan at the time. In 2023, her 5-carat ruby sold for 1.5 million yuan.

5. 红宝石投资收藏建议
Ruby Investment Collection Suggestions

(1) 权威证书
Authority Certificate

购买时要求提供权威宝石实验室的证书，证明红宝石的自然状态和品质。

At the time of purchase, you should request the seller to provide a certificate from the gem laboratory, which proves the natural state and quality of the ruby.

(2) 高品质红宝石
High Quality Ruby

优先选择高品质、颜色鲜艳、透明度好的红宝石。寻找未经热处理的红宝石，因为它们更稀有且价值更高。

Priority to choose high quality, bright color, good transparency rubies. Look for rubies that have not been heat-treated, as they are rarer and more valuable.

(3) 市场趋势
Market Trends

了解国际市场价格波动，在投资前咨询名贵珠宝投资顾问或者专门的宝石公司。

Understand the price fluctuations in the international market, and consult luxury jewelry investment advisers or specialized gem companies before investment.

小雨老师经验分享
Experience Sharing from Ms. Xiaoyu

小雨老师最经典的那句话就是，世界上99%的珠宝都是不保值的，那么对于红宝石而言，我们一定要买高精尖的那一部分，怎么说呢？比如说你想要买一颗2克拉或者3克拉的红宝石，你一定要买高货，如果是擦边货就不行。拿现在的价格来讲，2克拉的如果只需要几万元，不到10万元买到的，那肯定不是精品。按照现在市场的行情，2克拉的精品货可以达到20万元人民币以上。然后，可能在五年前大家的印象里面，3克拉的精品货可能也就15万元人民币出头，现在3克拉的精品货价格在50万元人民币左右，所以说你如果五年前、三年前买的是高货，现在非常保值增值。

所谓的高货，它的6个维度一定要讲清楚，第一个是净度，红宝石由于它天生的裂隙发育就比较明显，所以说如果它微微有一点瑕疵，或者在腰棱上有点儿矿缺，其实我们都是能够接受的，因为毕竟它是一个天然的东西；在颜色上最好达到鸽血红；在重量上想要保持增值，至少要2克拉以上；切工上至少要整个宝石是亮亮的，不是很暗的，不是那种发黑的；优化上，最好是天然无烧的，如果是大克拉天然有烧的也是可以的；证书的话，GRS、古柏林、SSEF都很好。如果是想要投资收藏作为资产配置，那么同样的高精尖的货，要5克拉以上，目前价格在250万元人民币左右了。

The most classic saying of Teacher Xiaoyu is that: 99% of the jewelry in the world is not valuable. So for ruby, we must buy the sophisticated part. How to understand? For example, if you want to buy a 2 or 3 carat ruby, you must buy high-end product. If it is a just-so-so stone, it cannot be selected. Take the current price for example, if the 2 carat only needs tens of thousands of yuan, less than 100,000 yuan to buy, it is certainly not boutique. According to the current market situation, 2 carats of fine goods can reach more than 200,000 yuan. Then, maybe in the impression of five years ago, the 3 carat fine goods may also be more than 150,000 yuan, and now the 3 carat fine goods price is about 500,000 yuan. So if you bought high-end goods five years ago or three years ago, it is now very preservation and appreciation.

The so-called high-end goods, you must be clear about its 6 dimensions. The first is clarity, ruby, asts natural crack development is more obvious, so if it has a slight flaw, or a little ore shortage on the waist edge, in fact it is acceptable for us. Because after all, it is a natural thing; It is best to achieve pigeon blood in color. In order to maintain the increase value in weight, choose the ones of at least 2 carats. At least the whole stone should be bright, not very dark, not black. In terms of optimization, it is best to be natural without burning, if it is a large carat natural burning will also be acceptable. For certificates, GRS, Gubelin, SSEF are all good. If you want to invest in collection as an asset allocation, then similarly choose the sophisticated goods, which are more than 5 carats, whose current price is about 2.5 million yuan.

The Lotus Color Spinel

第三章
03 Chapter Three

尖晶石
Spinel

一、尖晶石的基本特征
Basic Characteristics of Spinel

尖晶石，英文名称为"spinel"，源自拉丁语"spina"，意为"荆棘"，尖晶石(spinel)成分为镁铝氧化物。尖晶石与红宝石常共生，俗称"大红宝"。在国际市场上非常好流通，也是拍卖场上非常受欢迎的名贵珠宝之一，很多国际知名珠宝大品牌都会用到尖晶石。作为唯一能跟红宝石媲美的红色系宝石，未来非常有投资价值。红色的尖晶石极易与红宝石混淆，如镶嵌在英国皇冠上的"黑王子红宝石"和"铁木尔红宝石（timur ruby）"实际上都是尖晶石。其基本特征包括稀有性、高透明度、颜色丰富，是收藏新贵。尖晶石主要产地包括缅甸、斯里兰卡、尼日利亚、坦桑尼亚等地。

Spinel, from the Latin "spina", meaning "thorn". Spinel is composed of magnesium aluminum oxide. Spinel and ruby are often symbiotic, commonly known as "great red treasure". It is very good in circulation in the international market, and is also one of the most popular precious jewelry on the auction floor, and many internationally renowned jewelry brands will use spinel. As the only red gemstone that can be compared with ruby, it will be a very valuable investment in the future. Red spinel is easily confused with rubies, such as the "Black Prince Ruby" and "Timur Ruiby" embedded in the British crown are actually spinel. Its basic characteristics include rarity, high transparency, rich colors, which is a collection of new rich. The main production areas of spinel include Myanmar, Sri Lanka, Nigeria, Tanzania and other places.

小雨老师与原产地商人进行宝石交易

小雨老师宝石交易

马亨盖尖晶石戒指

Basic Characteristics of Spinel

Spinel is highly sought after in the international market and is considered a precious gemstone. Many renowned international jewelry brands incorporate spinel in their designs.

Spinel

二、如何鉴赏和投资收藏尖晶石
How to Appreciate and Invest in the Spinel Collection

我们主要从以下 7 个维度去鉴赏和投资一颗尖晶石，我在每一个鉴赏指标分析里都给出了目前的市场情况以及相应指标的价格影响，也参考了一些拍卖价格，大家可以参考。

We mainly appreciate and invest a spinel from the following seven aspects. In each appreciation index analysis, I have given the current market situation and the price impact of the corresponding indicators, and also referred to some auction prices for your reference.

挑选时，要把颜色排在第一位，颜色饱和度越高越好，价格越高。其中，红色调且强荧光的绝地尖晶石，还有马亨盖的红色系列尖晶石，在市场上都非常受欢迎。蓝色系列，钴尖晶石最有价值，只有一些顶级珠宝大牌才会用到钴尖晶石。由于尖晶石的产地特别多，而且不同产地颜色不一样，我们在收藏的时候尽量选择颜色浓郁一点的。

When choosing, the color should be ranked first, the higher the color saturation, the better, and the higher the price. Among of which, the red tone and strong fluorescent Jedi knight spinel, as well as Mahenge's red series spinel, are very popular in the market. Blue series, cobalt spinel is the most valuable, only some top jewelry brands will use cobalt spinel. As the origin of spinel is too much, and the color of different origin is not the same, we will try to choose a strong color when collecting.

小雨老师经验分享
Experience Sharing from Ms. Xiaoyu

尖晶石的颜色占到整个总价值的 50% 以上，一颗绝地尖晶石或者一颗马亨盖尖晶石，比一颗普通颜色的尖晶石可能要高出 10 倍的价格。在其他指标相同的情况下，一颗 2 克拉的普通红色尖晶石需要 2 万元人民币，但如果达到绝地尖晶石的颜色，可能就需要 20 万。一颗 10 克拉的马亨盖高品质红色系列的尖晶石，拍卖价格在 180 万元人民币左右。一颗 7 克拉的高品质钴尖晶石，价格在 400 万元人民币左右。

The color of the spinel accounts for more than 50% of the total value, and a Jedi spinel or a Mahenge spinel can be 10 times more expensive than a common color spinel. In the case of other indicators being the same, a 2-carat ordinary red spinel needs 20,000 yuan, but if it reaches the color of Jedi spinel, it may need 200,000 yuan. For a 10-carat Mahenge high-quality red series spinel, the auction price is about 1.8 million yuan. For a 7 carat high quality cobalt spinel, the price is about 4 million yuan.

（一）尖晶石的鉴赏
Appreciation of Spinel

1. 颜色
Color

是指尖晶石有哪些颜色。
It refers to the color of the spinel.

红色尖晶石
Red Spinel

红色尖晶石最出名的有两种，一个是来自缅甸的绝地尖晶石，一个是马亨盖尖晶石。红色是一个系列，并不是说只有红色。有偏玫色一点的红玫色，也有偏霓虹感强一点的正红色，也有偏粉一点的热粉色，总体而言，红色系列非常受欢迎。出自缅甸的尖晶石，有两种颜色很受欢迎，一是正红色缅甸尖晶石，另一个是带着霓虹感，叫做绝地尖晶石。坦桑尼亚的尖晶石以马亨盖尖晶石为主，它的颜色主要有三种，一是正红色，还有那种奶奶的带有一点粉红色叫做奶盖尖晶石，还有热粉色，都非常受藏家喜欢。很多人说尖晶石和红宝石从颜色上分不清。其实我们从火彩上就可以区分尖晶石和红宝石，尖晶石的火彩很闪，有一定的刚性，而红宝石的火彩比较温和，不会那么亮晶晶。

玻璃体马亨盖尖晶石

There are two most famous types of red spinel. The first one is the Jedi spinel from Myanmar, and the other is the Mahenge spinel. Red is a series of colors, not just red this single color. There are rosy red colors that are a little more rosy, there are positive red colors that are a little more neon, and there are hot pinky red colors that are a little more pink. And overall, the red series is very popular. Spinel from Myanmar, there are two colors are very popular, one is red Myanmar spinel, the other is with a neon sense, called Jedi spinel. The majority of Tanzania spinel are Mahenge spinel, its color is mainly three kinds. One is positive red, one is the milky with a little pink called milk cap spinel, and one is hot pink. The y are very popular among collectors. Many people say that the color of spinel and ruby is indistinguishable. In fact, we can distinguish between spinel and ruby from the fire color, the fire color of spinel is very bright, and there is a certain rigidity. And the fire color of ruby is relatively mild, not so shiny.

蓝色尖晶石
Blue Spinel

最出名的就是钴尖晶石，这种蓝色不是蓝宝石那种浓郁的皇家蓝或者矢车菊蓝色。钴尖晶石根据产地，颜色会不一样，来自越南的钴尖晶石，我们又叫它蓝小妖，真的像蓝色妖姬玫瑰花一样，很让人入迷的颜色。而坦桑尼亚的钴尖晶石，和斯里兰卡的钴尖晶石，会带有一点湛青色。提到越南这个产地，除了它的蓝色钴尖晶石很出名以外，还有一些越南的紫色尖晶石、莲花色的尖晶石、热粉色尖晶石也都特别漂亮。

钴尖晶石

The most famous is cobalt spinel, which is not the rich royal blue or cornflower blue of sapphire. Cobalt spinel, according to the different origin, the color will be different. The cobalt spinel from Vietnam, we also call it the blue demon, it is really like the blue demon rose, a very fascinating color. And the cobalt spinel from Tanzania, and the cobalt spinel from Sri Lanka, which will have a little bit of cham blue. When it comes to the origin of Vietnam, in addition to its famous blue cobalt spinel, there are also some Vietnamese purple spinel, lotus color spinel, hot pink spinel. They are also particularly beautiful.

紫色尖晶石
Purple Spinel

阿富汗的帕米尔高原上的紫色尖晶石，我们又叫它鸢尾尖晶石，让人想到凡·高的画作《鸢尾花》。

The purple spinel in Afghanistan's Pamir plateau, which we also call iris spinel, reminds us of Van Gogh's painting Iris.

2. 净度
Clarity

是指尖晶石的内外的纯净程度，是否含有包裹体。
It refers to the inner and outer purity of the spinel, whether it contains inclusions.

全净尖晶石

包裹体的大小、数量和位置
Size, Quantity and Location of Inclusions

包裹体越大、数量越多，对宝石的净度影响就越大。特别是当包裹体位于台面部位时，对宝石净度的影响更为显著。
The larger and more numerous the inclusions, the greater the impact on the clarity of the stone. Especially when the inclusion is located in the mesa, the influence on the clarity of the gem is more significant.

包裹体的类型和对比度
Type and Contrast of Inclusions

包裹体的折射率和颜色与尖晶石的差异也会影响宝石的净度。如果包裹体的折射率和颜色与尖晶石相差较大，对宝石的净度影响会更明显。
Differences in refraction and color of inclusions from spinel also affect the clarity of the gem. If the refractive index and color of the inclusion are different from that of spinel, the effect on the clarity of the gem will be more obvious.

未愈合的裂隙
Unhealed Cracks

如果宝石内部有未愈合的裂隙，将降低宝石的耐久性，容易受到损伤。
If there are unhealed cracks inside the stone, it will reduce the durability of the stone and make it vulnerable to damage.

肉眼可见开放裂隙与包裹体的尖晶石

尖晶石通常情况下净度都不会太好，因为它内部一般会含有附晶和一些小的愈合裂隙，我们在选购的时候尽量选瑕疵在底部或者在侧棱边上的，肉眼在社交距离不能发现，不影响它的美观即可。在台面上最好不要出现很明显的包裹体，总之，瑕疵越少越好。针对宝石而言，如果这颗宝石越大，我们对它的净度要求是越低的。如果宝石很大一颗，颜色又那么好，再要求它净度是全净的话，其实很难满足的。所以说如果满足了克拉大、颜色好、切工好，只是有一点瑕疵，我们在市场上都是可以接受的。

Under normal circumstances, the clarity of spinel is not too good, because it generally contains crystals and some small healing cracks inside. When doing shopping, we try to choose those whose defects are at the bottom or on the edge of the side. It is hard to be found by the naked eye in the social distance, and does not affect its beauty. It is best not to appear very obvious inclusions on the platform, in short, the fewer defects the better. For gemstones, if the gemstone is larger, we have lower clarity requirements for it. If the stone is very big, the color is so good, a nd then it is required to be completely clean, it is actually difficult to meet. So if it meets the criteria of big carats, good color, good cut, and there is only a little flaw, then it is acceptable in the market.

小雨老师经验分享
Experience Sharing from Ms. Xiaoyu

尖晶石的净度其实跟大多数宝石一样，如果这颗宝石越大，我们对它的净度要求应该稍微低一点儿，但如果这个宝石比较小，那我们对它的净度就可以要求高一点儿。但是也有例外，绝地尖晶石本身就长不大，而且它的裂隙发育本身就比较强，所以针对这种特殊的宝石品种，只要他的霓虹感很强，其他净度差一点，不会太多影响它的价格。但马亨盖尖晶石，是可以长得比较大的，所以我们对它的净度要求会更高一些。比如说一颗10克拉的高品质马亨盖尖晶石，如果是那种干净玻璃体的，大概要180万元人民币左右，但如果它净度稍微差了一么一点点，价格可能就会在120万元人民币左右。

The clarity of spinel is actually like most gemstones, if the gemstone is larger, we should have a slightly lower clarity requirement, but if the gemstone is relatively small, then we can have a higher clarity requirement. But there are exceptions, the Jedi spinel itself is not long, and its fissure development itself is relatively strong, so for this special gemstone variety, as long as his neon sense is strong, other clarity is less, then it will not affect its price too much. But the Mahenge spinel can grow relatively large, so we will have higher clarity requirements for it. For example, a 10 carat high-quality Mahenge spinel, if it is the kind of clean vitreous body, the price will be about 1.8 million yuan. But if its clarity is a little bit worse, the price may be about 1.2 million yuan.

3. 切工
Cut

是指宝石外表切割的对称度和工整度。
It refers to the symmetry and smoothness of the cut of the gem's appearance.

尖晶石切割的好坏会影响它在台面上呈现出来的火彩,那种亮晶晶的为最佳。切割得越好,它的反火越好,那么整个宝石就会越亮。如果切割得不好,比如说有漏底,或者过厚。漏底就会出现空窗效应,中间的颜色就不浓郁;但如果是过厚的话,它会发黑,而且台面的比例就会变小。明明是一颗5克拉的看上去像3克拉,从经济价值上就不划算,所以我们尽量选择切割比例舒展的尖晶石。

The quality of spinel cutting will affect the fire color it presents on the platform, and the shiny one is the best. The better the cut, the better its backfire, then the brighter the whole gem will be. If it's not cut well, for example it has a leaky bottom, or it's too thick. Leaky bottom will lead to empty window effect, the middle color is not rich; But if it is too thick, it will appear black, and the proportion of the platform will be smaller. Obviously one spinel is a 5 carat but looks like 3 carat, from the economic value it is not cost-effective, so we try to choose the kind of spinel whose cut proportion is stretch spinel.

漏底尖晶石改切

薄底厚底尖晶石

小雨老师经验分享
Experience Sharing from Ms. Xiaoyu

切割最主要的就是不要选有漏底的,以及切得过厚的,尽量选择切割的比例比较周正的,打个比方,一颗5克拉的高品质的尖晶石,如果它切得有点漏底的话,可能就从50万元人民币变成30万元人民币。

The most important thing to cut is not to choose the bottom of the leak, and cut too thick, as far as possible to cut the proportion of the more positive, for example, a 5 carat high-quality spinel, if he cut a little bit of the leak, it may be from 500,000 yuan to 300,000 yuan.

小雨老师经验分享
Experience Sharing from Ms. Xiaoyu

目前10克拉的高品质红宝石价格是在2500多万元人民币,而10克拉的马亨盖尖晶石目前价格在300万元人民币左右,所以说红色的尖晶石是非常有潜力的,尤其是大克拉的。

At present, the price of 10 carat high-quality ruby is more than 25 million yuan, and the current price of 10 carat Mahenge spinel is about 3 million yuan, so the red spinel is very promising, especially the large carat.

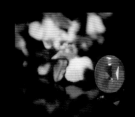

13.23克拉 天然"缅甸"未经加热尖晶石戒指
苏富比香港
成交价:HKD1,397,000

4. 克拉重量
Carat Weight

不同产地的尖晶石,由于品种原因,我们对它的克拉重量应该分别去看待。比如越南的钴尖晶石,就是重量特别小,1克拉以上的钴尖晶石就已经很难得了,3克拉以上的越南产地的钴尖晶石,那肯定值得收藏了。纳米亚和辛巴的绝地尖晶石,它也是本身的大小就很小,所以切割出来一般不会超过1克拉,1克拉以上就很好了,如果有一颗3克拉的绝地尖晶石,绝对是值得收藏的。那像坦桑尼亚的马亨盖的尖晶石,还有阿富汗帕米尔高原的紫色尖晶石,就要大一些,我们尽量选择大克拉的。缅甸的红色尖晶石和马亨盖尖晶石个体大,是红宝石最好的平替品,而且红色尖晶石真的非常漂亮,刚性和火彩都很好,比红宝石还要漂亮。

Spinel of different origin, due to variety reasons, we should treat its carat weight separately. For example, Vietnam's cobalt spinel, the weight is particularly small. Those cobalt spinels more than 1 carat are already difficult to get, and more than 3 carats in Vietnam will be definitely worth collecting. Cobalt spinel in Nania and Simba Jedi spinel, are also very small in its size, so cut out generally will not more than 1 carat. A cobalt spinel more than 1 carat is very good, and if there is a 3 carat Jedi spinel, it is definitely worth collecting. The spinels, like the ones from Mahenge in Tanzania, and the purple ones from the Pamir plateau in Afghanistan, are generally bigger. So we need to choose the one that are larger carats. Myanmar's red spinel and Mahenge spinel are large individuals, and are the best flat ruby, and red spinel is really very beautiful, its rigid and fire color is very good, and more beautiful than ruby.

5. 产地
Origin

尖晶石的产地跟颜色是息息相关的，也决定了尖晶石的价值，比如缅甸的尖晶石主要是以正红色居多，缅甸的绝地尖晶石很值得投资。越南的尖晶石主要以蓝色的钴晶石、紫色尖晶石，还有莲花色的尖晶石著称，然后马亨盖的尖晶石是以丝绒感奶奶的，还有玻璃体的亮亮的著称。绝地的尖晶石主要以霓虹的色感著称。尖晶石的产地跟它的颜色是挂钩的，也决定了尖晶石的价值。

The origin of spinel is linked to the color, which also determines the value of spinel, such as Myanmar's spinel is mainly positive red, and Myanmar's Jedi spinel is worth investing in. Vietnam's spinel is mainly known for blue cobalt, purple spinel, and lotus colored spinel. And the Mahenge spinel is known for its velvet, as well as the bright glass body. Jedi spinels are known primarily for their neon color. The origin of spinel is linked to its color, which also determines the value of spinel.

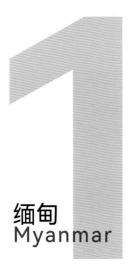

1 缅甸 Myanmar

绝地尖晶石

缅甸尖晶石以强荧光霓虹色的绝地尖晶石著称。缅甸尖晶石的特点包括其丰富的颜色，常见的有红色、橙色等。尖晶石因其稀有、透明度高、颜色艳丽而成为珍贵的宝石。产地主要分布在缅甸，能产出称得上绝地武士的尖晶石，主要有两个地方：缅甸克钦邦的纳米亚和缅甸抹谷的曼辛。纳米亚是矿区的名字，是绝地这一名字的起源地，最早发现绝地的地方。2001年，GIA宝石学家文森特·帕迪乌首先在纳米亚发现这种尖晶石，随后，将其命名为"绝地尖晶石"。这种尖晶石具有鲜艳明亮霓虹感、电光感的粉调红色，颜色纯净，没有一点的暗色域颜色像极了《星球大战》中绝地们使用的光剑！

Myanmar spinel is known for its strong fluorescent neon color Jedi spinel. The characteristics of Myanmar spinel include its rich colors, normally including red, orange and so on. Spinel is a precious gemstone because of its rarity, high transparency and bright color. The production is mainly distributed in Myanmar, which can produce spinels called Jedi knights, and mainly in two places: Namya in Kachin State, Myanmar, and Mansing in Mogok, Myanmar. Nania is the name of the mine, the origin of the name Jedi, where the Jedi were first discovered. In 2001, GIA gemologist Vincent Pardiu first discovered this spinel in Nania, and subsequently named it "Jedi spinel". This spinel has a bright neon, electric light pink tone red, pure color, and no dark gamut color, just like the Star Wars Jedi used lightsabers!

2013年8月，发现缅甸的曼辛（Man Sin）矿区也有一个绝地尖晶石的矿源。绝地武士，最重要的就是荧光颜色，"绝地尖晶石（Jedi Spinels）"，指具有独特的霓虹（电光感、荧光感）的粉—红色、没有暗域的尖晶石，并不是特指某个产地。相比其他颜色的尖晶石，绝地尖晶石的颗粒较小，品质优秀的绝地尖晶石，超过1克拉，已经非常难得，超过3克拉就算是很稀有了。

In August 2013, a source of Jedi spinel was discovered in the Man Sin mine in Myanmar. Jedi knight, whose most important is the fluorescent color. "Jedi knight Spinels", refers to a unique neon sense (electric light, fluorescent sense) pink - red, no dark area of spinel, and it is not a specific origin. Compared with other colors of spinel, Jedi knight particles are small. A good quality Jedi knight, weight more than 1 carat is already very rare, if weight more than 3 carats, it is very previous.

越南 Vietnam

越南的尖晶石以蓝色的钴尖晶石著称。越南尖晶石包括各种色调的红色、蓝色、紫色、橙色等。越南的粉色尖晶石通常明度较高，一个字总结越南的尖晶石就是"仙"——仙紫色尖晶石，热粉色尖晶石，莲花色尖晶石。越南最出名的钴尖晶石，市面上又叫它"蓝小妖"，但是颗粒很小，大于1克拉都算是很大的了。小雨老师特别提醒：钴尖晶石在国内的价格远远被低估，其实在国际上非常有投资价值，我在美国纽约见的一些老牌犹太珠宝商，一颗大克拉的钴尖晶石可以出到很高的价格，他们最好的钴尖晶石都会卖给大牌的珠宝商，比如梵克雅宝，还有一些市面上没听过，只在贵族富豪中交易的小众高端品牌。

Vietnam's spinel is known for its blue cobalt spinel. Vietnamese spinel includes various shades of red, blue, purple, orange, etc. Vietnam's pink spinel is usually bright. One word to sum up Vietnam's spinel is "fairy" - fairy purple spinel, hot pink spinel, lotus color spinel. Vietnam's most famous cobalt spinel is the so-called "blue demon" in the market, but the particles are very small. The one weight more than 1 carat is considered very large. Special reminds from Ms. Xiaoyu: Cobalt spinel in the domestic price is far undervalued. In fact, in the international its very investment value. I met some old Jewish jewelers in New York. A large carat cobalt spinel can be sold with a very high price. Their best cobalt spinel will be sold to the big jewelers, such as Van Cleef & Arpels, and there are some niche high-end brand, which have not heard of on the market, but onyl traded only among the aristocratic rich.Ω

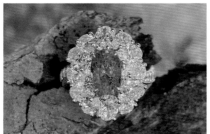

越南钴尖晶石

坦桑尼亚 Tanzania

马亨盖尖晶石

马亨盖（Mahenge）尖晶石产自坦桑尼亚，是一种带有霓虹观感的粉色尖晶石。优质的马亨盖尖晶石呈艳粉色，颜色饱和度高并略带丝绒感。20世纪80年代曾于莫罗戈罗省（Morogoro）的马通博（Matombo）和马亨盖（Mahenge）附近发现尖晶石。马亨盖尖晶石颜色由粉色到红色，聚集了热粉色、橘粉色、霓虹粉、紫粉色、火焰红等色彩，具有珍贵的霓虹效应，拥有顶级的电光色、极少见的丝绒光带来的特殊晶质之美，是马亨盖尖晶石最具魅力之处。其中最为昂贵的是具备"Vibrant"的艳粉或红粉色。

Mahenge crystal, from Tanzania, is a pink spinel with a neon look. High quality Mahenge pointy crystals are bright pink, highly saturated and slightly velvety. In the 1980s, spinels were found near Matombo and Mahenge in Morogoro province. Mahenge spinel colors are from pink to red, gathered hot pink, orange pink, neon powder, purple pink, flame red and other colors, with precious neon effect. It has top electric color, rare velvet light which brought special crystalline beauty. And that is Mahenge spinel's the most attractive place. The most expensive of these is a Vibrant pink or red pink with a "Vibrant" color.

4 阿富汗 Afghanistan

帕米尔紫色尖晶石

以帕米尔高原的紫色鸢尾尖晶石（Iris Purple Spinel）著称，这些颜色瑰丽的紫色尖晶石，具有优秀的透明度、颜色饱和度高。与大多数宝石矿区相比，命运坎坷的阿富汗，战乱频发；且比起其他矿区，更加山高路险。对于专业采矿集团、宝石探险家们，想要到达阿富汗的矿区，困难重重，开采难度极大。在阿富汗发现的罕见、珍稀、美丽的宝石，却一直惊艳着国际宝石市场。历史悠久的巴达赫尚山谷，近年来，发现了一种色泽瑰丽、极具魅力、令人感官上无比愉悦的紫色尖晶石。

Known for the Pamir Iris Purple Spinel, these magnificent purple spinels have excellent transparency and high color saturation. Compared with most gem mining areas, Afghanistan, which has a rough fate, is prone to frequent wars, and more mountainous than other mining areas. For professional mining groups and gem explorers, it is difficult to reach the mining areas of Afghanistan. Rare, precious and beautiful gems found in Afghanistan have always amazed the international gem market. In the historic Badakhshan Valley, in recent years, a purple spinel, with a magnificent color, great charm and great pleasure to the senses has been discovered.

小雨老师经验分享
Experience Sharing from Ms. Xiaoyu

尖晶石的产地非常多，不同的产地有自己比较著名的宝石，比如说缅甸产地，以纳米亚和辛巴的绝地尖晶石为主，缅甸产地的红色尖晶石也非常有市场价值。阿富汗的帕米尔高原就是紫色尖晶石。坦桑尼亚就是马亨盖尖晶石。越南的尖晶石以蓝色的钴尖晶石为主，还有一些紫色、粉色、莲花色的尖晶石都非常有市场价值。目前在欧美市场上一颗7克拉以上的越南钴尖晶石大概市场价位是在500万元人民币左右。

There are many sources of spinel, and different sources have their own well-known gems, such as Myanmar origin, Jedi spinel from Nania and Simba, and red spinel from Myanmar origin are also very valuable in the market. The Pamirs of Afghanistan are famous for purple spinels. Tanzania is famous for the Mahenge spinel. Vietnam's spinel is dominated by blue cobalt spinel, and some purple, pink, lotus colored spinel, they are very valuable in the market. At present, in the European and American markets, the market price of a Vietnamese cobalt spinel above 7 carats is about 5 million yuan.

6.优化
Optimize

尖晶石其实很少会有优化处理这种情况，因为如果这个尖晶石是处理的，在证书上也会写清楚，但它的价格就会大打折扣。尖晶石有烧的，属于优化，但是市面上很少出现，通过证书也能看出来，从保值增值的角度是不会购买的。

In fact, spinel will rarely be optimized for this situation, because if the spinel is treated, it will be clearly written on the certificate, but its price will be greatly discounted. Spinel has burned, belongs to optimization, but rarely appears on the market, through the certificate can also be seen, from the perspective of preservation and appreciation is not purchased.

小雨老师经验分享
Experience Sharing from Ms. Xiaoyu

因为市面上如果这颗尖晶石是优化过的，我们肯定不能收藏，所以不做价格评估。

Because if this spinel is optimized on the market, we certainly cannot collect it, so we do not do price assessment.

7. 证书
Certificates

不同的证书代表不同的含金量
Different certificates represent different gold content.

01 GUILD

吉尔德的证书，来自美国吉尔德实验室。其评价标准中，主要做祖母绿的证书，也有其他彩宝的证书。GUILD也会标注产地、有无加热、颜色。这一颗宝石是来自于缅甸的话，它会出现绝地尖晶石的字样。

Guild's certificate, from Guild Laboratories in the United States. Among its evaluation criteria, it mainly does the certificate of emerald, and also has the certificate of other color treasure. Guild also notes origin, heating, and color. If this gem came from Myanmar, it would have the word 'Jedi spinel'.

GUILD马亨盖尖晶石证书

02 GRS

来自瑞士的宝石实验室，主要做彩色宝石的鉴定，含金量也很高，也会出现在拍卖场上，如果宝石品质足够好，还会有铂金证书。GRS会标明产地、有无加热以及颜色的评级。GRS证书会出马亨盖尖晶石，它会有vibrant的评价，相当于对这颗尖晶石非常的认可。

From the gemstone laboratory in Switzerland, mainly to do the identification of colored gems. Its gold content is also high, and will also appear in the auction. If the gemstone quality is good enough, there will be a platinum certificate. GRS will indicate the origin, heating, and the color rating. The GRS certificate will be issued to the Mahenge spinel, and it will have a vibrant evaluation, which is equivalent to the high recognition of this spinel.

GRS vibrant尖晶石证书

03 AIGS

亚洲宝石学院是一家总部位于泰国曼谷的宝石学院，对颜色达到一定高饱和度的，会给出绝地尖晶石的评级。

The Asian Gemological Institute, a gemological institute based in Bangkok, Thailand, rates Jedi spinels for colors that reach a certain high saturation.

AIGS绝地武士尖晶石戒指

AIGS绝地尖晶石证书

小雨老师经验分享
Experience Sharing from Ms. Xiaoyu

尖晶石的颜色占到总价值的50%以上，一颗绝地尖晶石或者一颗马亨盖的尖晶石颜色，比一颗普通颜色的尖晶石可能要高出10倍的价格。在其他指标相同的情况下，一颗2克拉的普通红色尖晶石需要2万元人民币，但考虑证书对价格的影响：目前尖晶石的证书主要是GRS，因为好的尖晶石它会有vibrant评级，还可以出到铂金证书，这样含金量就比较高，在市场上的说服力也是非常高的。如果达到绝地尖晶石的颜色，可能就需要20万元人民币。一颗10克拉的马亨盖高品质红色尖晶石，拍卖价格在180万元人民币左右。一颗7克拉的高品质钴尖晶石，价格在400万元人民币左右。

Spinel color accounts for more than 50% of the total value, and a Jedi spinel or a Mahenge spinel color can be 10 times more expensive than a common color spinel. In the case of the same other indicators, a 2 carat ordinary red spinel costs 20,000 yuan. But consider the impact of the certificate on the price: at present, the certificate of spinel is mainly GRS, as a good spinel will have a vibrant rating, and it can also be issued to the platinum certificate. So the gold content is relatively high, and the persuasive force in the market will be also very high. If it reach the color of Jedi spinel, it may be priced at 200,000 yuan. A 10-carat Mahenge high-quality red spinel, the auction price is about 1.8 million yuan. A 7-carat high quality cobalt spinel at around 4 million yuan.

(二)尖晶石的投资收藏建议
Spinel Investment and Collection Suggestions

1.尖晶石涨价的原因
The Reason for the Price Rise of Spinel

就目前尖晶石矿产资源而言,全世界已发现的尖晶石矿点有1000多处,但宝石级尖晶石的产地却相当稀少。2023年,坦桑尼亚尖晶石在短短一年内实现了市场份额的飞跃增长,由2021年的5.69%增长到如今的17.72%。尖晶石的主要产区是缅甸和坦桑尼亚,缅甸的尖晶石最出名的是绝地尖晶石,颜色非常特别,但是它的晶体太小了,3克拉的绝地尖晶石都算是高品质了。坦桑尼亚的马亨盖尖晶石颜色漂亮,而且晶体比较大,视觉上非常抢眼。红色的马亨盖尖晶石看起来非常像红宝石,红宝石的产量日益减少,所以能代替它的马亨盖尖晶石在欧美市场的价格一路高升。目前高品质的红宝石10克拉的价格在2500万元人民币左右,而高品质的10克拉马亨盖尖晶石价格是300万元人民币左右,他们有10倍的价差,所以还有很大的空间。还有钴尖晶石的价值一直被国内人低估,而在欧美国家钴尖晶石真的是一颗难求。它拥有良好的透明度和钴蓝色,其价值也可以是同类蓝宝石的10倍左右,但产量特别稀少,供给关系紧缺,一颗5到10克拉的钴尖晶石真的太值得投资了。给大家看看近几年的尖晶石的产量占比情况。

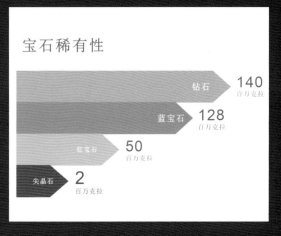

缅甸尖晶石价格趋势(1975—2022)
以美元结算
Myanmar Spinel Price Trend (1975-2022)
is Settled in US Dollars

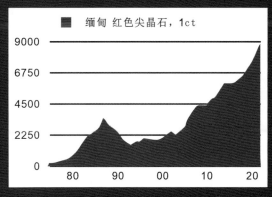

As far as spinel mineral resources are concerned, there are more than 1,000 spinel mineral sites that have been found in the world, but the origin of gem-grade spinel is quite rare. In 2023, Tanzanian spinel achieved a leap in market share in just one year, increasing from 5.69% in 2021 to 17.72% today. The main producing areas of spinel are Myanmar and Tanzania, Myanmar's spinel is most famous for Jedi spinel, whose color is very special, but its crystal is too small. A 3 carat Jedi spinel is considered to be very high quality. Tanzania's Mahenge spinel color is beautiful, and the crystal is relatively large, and visually very eye-catching. The red Mahenge spinel looks very Mahenge spinel looks very much like ruby, and the production of ruby is decreasing, so the price of Mahenge spinel that can replace it is soaring in the European and American markets. At present, the price of high-quality ruby 10 carat is about 25 million yuan, and the price of high-quality 10 carat Mahenge spinel is about 3 million yuan. There are 10 times of price difference, so there is still a lot of space for price increase. There is also the value of cobalt spinel has been underestimated by the domestic people, and in Europe and the United States cobalt spinel is really hard to find even one piece. It has good transparency and cobalt blue, and its value can also be about 10 times that of similar sapphires. But the production is particularly rare, and the supply is in shortage. A 5 to 10 carat cobalt spinel is really worth investing in. Let's take a look at the proportion of spinel production in recent years.

2. 尖晶石近几年的大约价格涨幅
（以高品质马亨盖尖晶石为例）

Approximate Price Increase of Spinel in Recent Years
(Taking High Quality Mahenge Spinel as an Example)

重量（克拉） Weight(Carat)	2.00-3.00克拉 （元人民币/克拉） 2.00-3.00ct（CNY/ct）	5.00克拉 （元人民币/克拉） 5.00ct（CNY/ct）	10.00克拉 （元人民币/克拉） 10.00ct（CNY/ct）	20.00克拉 （元人民币/克拉） 20.00ct（CNY/ct）
2019 年	15,000	50,000	80,000	130,000
2020 年	20,000	60,000	120,000	150,000
2021 年	30,000	80,000	170,000	200,000
2022 年	40,000	100,000	200,000	250,000
2023 年	50,000	120,000	220,000	300,000
2024 年	60,000	150,000	250,000	350,000

3. 尖晶石投资案例
Spinel Investment Case

2016年尖晶石在美国的图桑珠宝展、拉斯韦加斯珠宝展、瑞士日内瓦珠宝展上就非常地受欢迎了，但是在国内的市场还没有火爆起来。真正火爆起来是从2019年开始的，当时有个做医疗投资的朋友让我给他推荐好的投资型宝石，我就推荐了尖晶石，当时我建议他买5克拉或者10克拉的高品质尖晶石，他当时觉得价格很贵，到了2023年，他在香港卖出，将近赚了3倍，他就后悔买少了。

In 2016, spinel was very popular at the Tusson Jewelry Show in the United States, the Las Vegas Jewelry Show, and the Geneva Jewelry Show in Switzerland, but in the domestic market has not yet exploded. The really hot started from 2019, when a medical investment friend asked me to recommend for him about a good investment gemstone. I recommended spinel, and at that time I suggested that he buy 5 or 10 carats of high-quality spinel. He felt that the price is very expensive, but in 2023, he sold the spinel in Hong Kong, nearly three times the profit. And he regretted that he bought not enough back that time.

4.尖晶石拍卖价格
Spinel Auction Price

时间 Time	拍卖行 Auction House	拍品 Item	成交价 Closing Price
2022年2月 February, 2022	苏富比香港 Sotheby's Hong Kong	9.72克拉 天然"缅甸"未经加热尖晶石配钻石及沙弗莱石戒指 9.72 carat natural "Myanmar" unheated spinel with diamond and saffron ring	成交价 254,000 港元 Sale Price HKD 254,000
2023年5月 May, 2023	佳士得日内瓦 Christie's Geneva	20.83克拉枕形尖晶石戒指，产自坦桑尼亚，未经净度优化 20.83-carat pillow spinel ring, made in Tanzania, sold at a price not optimized for clarity	成交价 882,000法郎 Sale Price CHF 882,000
2018年秋拍 The Fall 2018 auction	佳士得香港 Christie's Hong Kong	20.11克拉椭圆形缅甸天然尖晶石胸针 20.11 carat oval Myanmar natural spinel brooch	成交价 1,812,500港元 Sale Price HKD 1,812,500

时间 Time	拍卖行 Auction House	拍品 Item	成交价 Closing Price
2018年秋拍 The Fall 2018 auction	佳士得香港 Christie's Hong Kong	9.22克拉枕形缅甸天然尖晶石戒 9.22 carat pillow shaped Myanmar natural spinel ring	成交价 1,125,000港元 Sale Price: HKD 1,125,000
2021年12月 December, 2021	邦瀚斯纽约 BonHams New York	8.21克拉变色钴尖晶石和钻石戒指，俄罗斯 8.21 carat color-changing cobalt spinel and diamond ring	成交价 20,312.50 美元 Sale Price: USD 20,312.50
2023年5月 May, 2023	佳士得香港 Christie's Hong Kong	4.03克拉坦桑尼亚钴尖晶石配钻石戒指 4.03 carat Tanzanian cobalt spinel with diamond ring	估价 800,000 — 1,200,000港元 Estimate: HKD800,000—1,200,000
2018年春拍 The Spring 2018 Auction	苏富比纽约 Sotheby's New York	31.03克拉天然坦桑尼亚粉红尖晶石配钻石戒指 31.03 carat natural Tanzanian pink spinel with diamond ring	成交价 935,528元人民币 Sale Price: RMB 935,528

5.尖晶石的投资收藏建议
Spinel Investment Collection Suggestions

(1) 权威证书
　　Authority Certificate

　　确保购买具有含金量高的鉴定机构认证的尖晶石。认证可以确保宝石的真实性和品质，并提供宝石的详细信息。

　　Be sure to purchase spinel certified by an accreditation body with a high gold content. Certification ensures the authenticity and quality of the stone and provides detailed information about the stone.

(2) 高品质石头
　　High Quality Stone

　　选择颜色鲜艳、透明度高、切割工艺精湛和克拉数大的石头，它们更有可能保持价值和增值潜力。留意稀有品种：某些尖晶石品种非常稀有，如马亨盖、绝地、钴尖晶石等。这些稀有品种通常具有更高的投资价值和收藏价值。

　　Choose stones with bright colors, high transparency, excellent cutting techniques and large carats, which are more likely to retain their value and increase their value potential. Watch out for rare varieties: Some spinel varieties are very rare, such as Mahenge, Jedi, cobalt spinel, etc. These rare varieties usually have higher investment value and collectible value.

(3) 市场趋势
　　Market Trends

　　定期了解尖晶石市场的动态和趋势，市场的供求关系、拍卖价格和珠宝大品牌都使用的品种，受欢迎程度都可能影响尖晶石的价值，最好咨询名贵珠宝投资顾问。

　　Regular understanding of spinel market dynamics and trends, market supply and demand, auction prices and jewelry brands are used varieties, popularity may affect the value of spinel. It is best to consult luxury jewelry investment advisers.

小雨老师经验分享
Experience Sharing from Ms. Xiaoyu

　　颜色是尖晶石最重要的，红色系列的尖晶石，比如缅甸的正红色尖晶石、坦桑尼亚的马亨盖尖晶石，都是红宝石最好的替代品，因为它的个头可以长到足够大，跟红宝石一样，而且它的整个透明度、火彩、刚性质感，会比红宝石表现得更好一些。坦桑尼亚的红色尖晶石，10克拉以上，目前的价格在200万元人民币左右，绝地尖晶石由于个头本身就比较小，2克拉以上就很有保值价值，目前的价格在20万元人民币以上。在欧美市场上，红色系列的尖晶石非常受欢迎。还有一些小众产地的尖晶石，比如越南钴尖晶石，即便是1克拉也很值得收藏，价格在10万元人民币左右，因为它本身就长不大。阿富汗帕米尔高原的紫色尖晶石，10克拉以上好品质，可能在100万元人民币左右，越大克拉越好，也是非常受欢迎的。在宝石的收藏圈里面，这些都是大家争相要去收藏的。

　　Color is the most important for spinel. Red series of spinel, such as Myanmar's positive red spinel, Tanzania's Mahenge spinel, are the best substitute for ruby. That is because their head can grow large enough to be the same as ruby, and its whole transparency, fire color, rigid texture, will be better than ruby. Tanzania's red spinel, more than 10 carats, its current price is about 2 million yuan. Jedi spinel, as its head itself is relatively small, one more than 2 carats is very valuable, its current price is more than 200,000 yuan. In the European and American markets, the red series of spinel is very popular. There are also some small sources of spinel, such as Vietnam cobalt spinel, which even 1 carat is also worth collecting. The price is about 100,000 yuan, as it itself is not big. The purple spinel of the Pamir Plateau of Afghanistan, one more than 10 carats of good quality, and its may be about 1 million yuan. The larger the carat the better, it is also very popular. In the gem collection circle, these are all competing to collect.

The Charm of **Padparadscha**

第四章
04 Chapter Four

帕帕拉恰
Padparadscha

一、帕帕拉恰的基本特征
Basic Characteristics of Padparadscha

帕帕拉恰英文名Padparadscha，帕帕拉恰是蓝宝石家族的一员，出身高贵，宝石硬度仅次于钻石，且产地仅有4个，产量也极为稀少，据考证帕帕拉恰的矿藏储量仅为蓝宝石的1%，矿源稀少程度堪比世界上产量最为稀少的鸽血红。帕帕拉恰指产自斯里兰卡的呈橙—粉色的蓝宝石。它在僧伽罗语中的意思是"莲花色"，因此得名帕帕拉恰蓝宝石。帕帕拉恰蓝宝石的颜色范围从浅色到中等色调，呈微粉红色到橙色，再到粉红和橙红色。一般来说，被称为帕帕拉恰的蓝宝石要求在整个宝石上，粉色和橙色的比例在30%至70%之间，并且不能有其他色彩的杂色。帕帕拉恰蓝宝石因其独特的颜色和稀有性而备受珠宝收藏家和爱好者的青睐。它的稀缺性和特殊的色调使其在市场上具有一定的收藏和投资价值。

Padparadscha is a member of the sapphire family, with noble birth. The gemstone hardness is second only to diamond, and there are only 4 origins, whose production is also very rare. According to research Padparadscha mineral reserves only account for 1% of sapphire. Its mineral scarcity is comparable to the world's rarest production of pigeon blood. Padparadscha refers to an orange-pink sapphire from Sri Lanka. It means "lotus color" in Sinhalese, hence getting the name Padparadscha sapphire. Padparadscha sapphires range in color from light to medium, with a slight pink to orange and to pink and orange-red. In general, the sapphire known as Padparadscha requires a ratio of pink to orange between 30% and 70% on the entire stone, and no other color variegations. Padparadscha sapphire is highly prized by jewelry collectors and enthusiasts because of its unique color and rarity. Its scarcity and special tone make it having a certain collection and investment value in the market.

Basic Characteristics of Padparadscha

帕帕拉恰裸石

帕帕拉恰戒指

小雨老师宝石交易

小雨老师与宝石商

帕帕拉恰套件

Basic Characteristics of Padparadscha

The color of Padparadscha sapphires is its most important price factor and one of the main reasons for its value and prestige.

Padparadscha

二、如何鉴赏和投资收藏帕帕拉恰
How to Appreciate and Invest in the Padparadscha Collection

我们主要从以下7个维度去鉴赏和投资一颗宝石，我在每一个鉴赏指标分析里都给出了目前的市场情况以及相应指标的价格影响，也参考了一些拍卖价格，大家可以参考。

We mainly appreciate and invest a gem from the following seven aspects. In each appreciation index analysis, I have given the current market situation and the price impact of the corresponding indicators, and also referred to some auction prices for your reference.

（一）帕帕拉恰的鉴赏
Appreciation of Padparadscha

1. 颜色
Color

是指帕帕拉恰有哪些颜色。
It refers to the color of Padparadscha

帕帕拉恰 帕帕拉恰蓝宝石的颜色是其最重要的价格因素，也是其珍贵和备受推崇的主要原因。

帕帕拉恰蓝宝石具有同时带有粉色和橙色色调的独特颜色，其颜色要求具有严格而微妙的比例，粉色和橙色的比例应在30%—70%之间，并且不能有其他杂色。只有同时满足这些条件，才能被认为是真正的帕帕拉恰蓝宝石。如果不满足这些条件之一，宝石将只被归类为粉色蓝宝石或橙色蓝宝石，而不是帕帕拉恰。较强烈、鲜艳的颜色通常被认为更为珍贵和稀有。不同的国际权威检测机构对帕帕拉恰颜色比例的定义略有不同。

The color of Padparadscha sapphire is its most important price factor and the main reason why it is precious and highly respected. Padparadscha sapphire has a unique color with pink and orange tones at the same time, and its color requirements have a strict and subtle ratio. The ratio of pink and orange should be between 30% and 70%, and there can be no other miscellaneous colors. Only when these conditions are met at the same time can it be considered a true Padparadscha sapphire. If one of these conditions is not met, the gemstone will only be classified as a pink sapphire or an orange sapphire, not a Padparadscha. Stronger, brighter colors are often considered more precious and rarer. Different international authoritative testing agencies have slightly different definitions of Padparadscha color ratio.

小雨老师经验分享
Experience Sharing from Ms. Xiaoyu

帕帕拉恰最重要的价格因素就是它的颜色，它的颜色一定是粉色跟橘色混合在一起的，有时候橘色调偏多一点，我们叫日落色。有时候它的粉色调偏多一点，我们叫日出色，这两种颜色在拍卖场上都非常受欢迎。目前一颗2克拉的高品质的帕帕拉恰，价格在10万元人民币左右。但如果是一颗高品质的5克拉的帕帕拉恰，价格目前是在上百万元人民币。

The most important price factor of Padparadscha is its color. Its color must be pink and orange mixed together, sometimes orange hue is a little more, we call sunset color. Sometimes it's a little bit more pink, which we call dayshine, and both colors are very popular at auction. At present, a 2 carat high quality Padparadscha, the price is about 100,000 yuan. But if it is a high-quality 5-carat Padparadscha, the price is currently will be millions of yuan.

帕帕拉恰 Padparadscha

粉橙色

Pastel Orange GRS将帕帕拉恰细分为"日出"和"日落"两种颜色类型，分别以粉色调和橙色调为主。当粉色和橙色各占50%时，这被认为是最优质的帕帕拉恰色。然而，这种比例非常罕见。因此，也可以接受粉色和橙色比例为55%和45%的宝石作为次优质的帕帕拉恰颜色。

Pastel Orange GRS subdivides Padparadscha into two color types, "sunrise" and "sunset," with pink and orange tones, respectively. When pink and orange are 50% each, this is considered the highest quality Padparadscha color. However, this proportion is very rare. Therefore, gems with pink and orange ratios of 55% and 45% can also be accepted as second-quality Padparadscha colors.

日落帕帕拉恰

日出帕帕拉恰

2.净度
Clarity

是指宝石的内外的纯净程度，是否含有包裹体。
This refers to the degree of purity of the stone inside and outside, whether it contains inclusions.

---

1. 包裹体的大小、数量和位置 / Size, Quantity and Location of Inclusions

包裹体越大、数量越多，对宝石的净度影响就越大。特别是当包裹体位于台面部位时，对宝石净度的影响更为显著。

The larger and more inclusions, the greater the impact on the clarity of the stone. Especially when the inclusion is located in the mesa, the influence on the clarity of the gem is more significant.

---

2. 包裹体的类型和对比度 / Type and Contrast of Inclusions

包裹体的颜色与宝石本身的颜色很相近的话，对宝石的净度影响不大。如果包裹体的折射率和颜色与红宝石相差较大，对宝石的净度影响会更明显。

If the color of the inclusion is very close to the color of the stone itself, it has little effect on the clarity of the gem. If the refractive index and color of the inclusion are significantly different from the ruby, the impact on the clarity of the gem will be more obvious.

---

3. 未愈合的裂缝 / An Open Fssure

如果宝石内部有未愈合的裂隙，将降低宝石的耐久性，容易受到损伤。

If there are unhealed cracks inside the gem, the durability of the stone will be reduced and it will be vulnerable to damage.

---

4. 丝状包裹体 / Filamentous Inclusions

少量丝状包裹体存在于宝石内部不会影响宝石的净度质量，反而可以改善宝石的外观。

The presence of a small amount of filamentous inclusions inside the stone does not affect the clarity quality of the stone, but can improve the appearance of the stone.

其实我们在实际生活中，肉眼或者社交距离看不到明显瑕疵，被认为是可接受的范围。然而，带有明显瑕疵并且影响宝石整体美观的宝石会被认为是劣质的。当然，完全无瑕疵的帕帕拉恰蓝宝石是最理想的，但这种品质的宝石非常罕见。大部分的帕帕拉恰蓝宝石都会有一些内含物或外部瑕疵存在，只要这些瑕疵不明显且不影响整体美感，仍然可以被认为是高品质的宝石。

明显瑕疵帕帕拉恰

In fact, in real life, we can not see obvious flaws with the naked eye or social distance, which is considered to be an acceptable range. However, stones with obvious flaws that affect the overall beauty of the stone are considered inferior. Of course, a completely flawless Padparadscha sapphire is ideal, but gemstones of this quality are very rare. Most Padparadscha sapphires will have some inclusions or external flaws present, as long as these flaws are not obvious and do not affect the overall beauty, it can still be considered a high quality stone.

净度优秀帕帕拉恰

小雨老师经验分享
Experience Sharing from Ms. Xiaoyu

只要从肉眼或者社交距离，不太影响美感，其实它的净度对价格的影响远远没有颜色那么大。

As long as with the naked eye or in social distance, there is no effect on the beauty, in fact, its clarity has far less impact on the price than the color.

3.切工
Cut

是指宝石外表切割的对称度和工整度。
It refers to the symmetry and fineness of the cut of the gem.

切工主要表现在对称性和抛光质量等方面。对称性问题指正侧面轮廓的对称性偏差、台面偏心、底尖偏心、亭部膨胀、刻面畸形和刻面尖点不尖等。抛光质量指的是宝石表面的抛光程度。红宝石的切割对其价格有显著影响。

由于帕帕拉恰蓝宝石的稀有性，宝石的形状通常由原石的形态决定。切割帕帕拉恰蓝宝石的目标是保持宝石的颜色和颜色强度，并尽可能减少内含物的影响。在切割帕帕拉恰蓝宝石时，切割师通常会选择那些最能展示宝石颜色和颜色强度的形状和切割方式。他们会努力保持帕帕拉恰蓝宝石的色彩，使其尽可能明亮和吸引人。同时，他们也会尽量减少内含物对宝石外观的影响，以确保宝石的整体品质。

The cutting is mainly manifested in symmetry and polishing quality. The symmetry problem refers to the symmetry deviation of the side profile, the eccentricity of the mesa, the eccentricity of the bottom tip, the bulge of the pavilion, the facet deformity and the non-cusp of the facet. Polishing quality refers to the degree of polishing of the gem's surface. The cutting of rubies has a significant effect on their price.

Due to the rarity of Padparadscha sapphires, the shape of the stone is usually determined by the shape of the raw stone. The goal of cutting Padparadscha sapphires is to preserve the color and color intensity of the stone and minimize the impact of inclusions. When cutting Padparadscha sapphires, cutters usually choose those shapes and cuts that best demonstrate the color and color intensity of the stone. They will strive to keep the color of the Padparadscha sapphire as bright and attractive as possible. At the same time, they will also minimize the impact of inclusions on the appearance of the stone to ensure the overall quality of the stone.

切工较差的帕帕拉恰　　切工优秀的帕帕拉恰

小雨老师经验分享
Experience Sharing from Ms. Xiaoyu

只要切割不影响宝石的整体的美观，切割对价值的影响就没有颜色那么大。

As long as the cut does not affect the overall beauty of the stone, the impact of the cut on the value is not as great as the color.

4.克拉重量
Carat Weight

一般来说，克拉重量越大的帕帕拉恰蓝宝石，价格会成倍增加。这是因为帕帕拉恰蓝宝石在自然界中非常罕见，较大的宝石更加稀有，因此更具价值。此外，大克拉的帕帕拉恰宝石也能展示其独特的颜色和光彩，更受人们的喜爱。

In general, the higher the carat weight of Padparadscha sapphire, the price will increase exponentially. This is because Padparadscha sapphires are very rare in nature, and larger stones are rarer and therefore more valuable. In addition, the large carat Padparadscha gemstone can also show its unique color and brilliance, and is more loved by people.

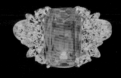

帕帕拉恰钻石戒指

小雨老师经验分享
Experience Sharing from Ms. Xiaoyu

3克拉以上的高品质帕帕拉恰目前的价格40万元人民币左右，而5克拉以上的高品质帕帕拉恰价格超过100万元人民币。

The current price of high quality Padparadscha of more than 3 carats is about 400,000 yuan, while the price of high quality Padparadscha of more than 5 carats is more than 1 million yuan.

5.产地
Place of Origin

是指帕帕拉恰的出产地，对价格影响挺大的。

It refers to where Padparadscha comes from, which has a big impact on the price.

1 斯里兰卡 / Sri Lanka

斯里兰卡是帕帕拉恰蓝宝石最知名的产地之一，之前，来自斯里兰卡的帕帕拉恰蓝宝石是最好的，也是最贵的，但是近几年发现，来自马达加斯加的帕帕拉恰蓝宝石，颜色、净度都非常好，所以他们之间的价格差距越来越小了。一些西方学者认为，只有在斯里兰卡产出的具有柔和粉橙色调的蓝宝石才能被称为真正的帕帕拉恰。然而，现在在国际上普遍承认的定义是，帕帕拉恰是指在标准日光下，具有明亮、浅到中等饱和度的刚玉，呈现出带有粉、橙两种色调的颜色。

斯里兰卡帕帕拉恰

Sri Lanka is one of the most well-known sources of Padparadscha sapphire. In the past, from Sri Lanka, Padparadscha sapphire is the best, but also the most expensive, but in recent years, it was found that gems from Madagascar, Padparadscha sapphire are very good, in terms of color and clarity. So the price gap between them is getting smaller and smaller. Some Western scholars believe that only sapphires with soft pinkish-orange tones produced in Sri Lanka can be called true Padparadscha. However, the internationally accepted definition now is that Padparadscha refers to corundum with bright, light to medium saturation under standard daylight, showing two shades of color with pink and orange.

2 马达加斯加 / Madagascar

马达加斯加产出的帕帕拉恰跟斯里兰卡的品质差不多。

马达加斯加帕帕拉恰戒指

Madagascar produces Padparadscha of similar quality to Sri Lanka.

小雨老师经验分享
Experience Sharing from Ms. Xiaoyu

其实帕帕拉恰这个小品种，产地的溢价没有那么大，斯里兰卡和马达加斯加现在都会产出高品质的帕帕拉恰，不会哪个产地的就贵一些，最重要的是整体漂亮才有价值。

In fact, for Padparadscha this small variety, the premium of origin is not so large, Sri Lanka and Madagascar are now both producing high-quality Padparadscha, which will not be more expensive. The most important thing is that the overall beauty is valuable.

6. 优化
Optimization

热处理
Heat Treatment

帕帕拉恰的热优化是一种常见的方法，它通过加热蓝宝石来改变颜色并使其稳定，经过热优化后的蓝宝石颜色均匀。热优化通常被市场接受，但需要在销售时明确注明。低温的热优化在实验室也能鉴定出来，在证书上也有标明是有烧。帕帕拉恰由于是色心致色，颜色通常不太稳定。我们在做GRS证书的时候，要求对帕帕拉恰的颜色做稳定性测试。即便测出来是稳定的，过段时间它可能就不稳定了，颜色会慢慢褪去，但是再过一段时间，颜色可能通过阳光照射等因素又慢慢恢复了。这个难题目前在各个实验室还没法攻破。当然，通过GRS稳定性测试的肯定更好。

Thermal optimization of Padparadscha is a common method, which changes the color and stabilizes the sapphire by heating it, and the color of the sapphire after thermal optimization is uniform. Thermal optimization is generally accepted by the market, but it needs to be clearly stated at the time of sale. The thermal optimization of low temperature can also be identified in the laboratory, and it is also indicated on the certificate that there is a burn. As Padparadscha is center-colored, the color is usually unstable. When we made the GRS certificate, we asked for stability testing of Padparadscha color. Even if it is stable, it may become unstable after a while, and the color will slowly fade, but after a while, the color may slowly return through factors such as sunlight exposure. It's a puzzle that's currently unsolvable in the LABS. Of course, passing the GRS stability test is definitely better.

颜色稳定性测试type1

小雨老师经验分享
Experience Sharing from Ms. Xiaoyu

如果一颗帕帕拉恰是优化过的，那么它的价值是远远没有天然的价值那么高的。打个比方，一颗5克拉的帕帕拉恰，如果是优化的，那么它的价值可能是20万元人民币，但如果是天然的，那么它的价值可能是上百万元人民币。针对优化的帕帕拉恰，目前市面上的接受度不高，不像蓝宝石跟红宝石，他们即便是优化的，国际上也是非常认可的，市场上也很认可。

If a Padparadscha is optimized, then its value is far less than the natural value. For example, a 5-carat Padparadscha, if optimized, may be worth 200,000 yuan, but if natural, it may be worth millions of yuan. For optimized Padparadscha, the current market acceptance is not high, unlike sapphire and ruby, even if they are optimized, it is also recognized well internationally and in the market.

Appreciation of Padparadscha

7. 证书
Certificate

不同的证书代表不同的含金量
Different certificates represent different gold content

01
GRS

该证书将帕帕拉恰细分为"日出"和"日落"两种类型。日出（sunrise）指饱和度中至强、以粉色调为主的（80%粉+20%橙）帕帕拉恰。日落（sunset）指饱和度中至强、以橙色调为主的（80%橙+20%粉）帕帕拉恰。而当粉色和橙色各占50%的时候，就造就了极佳的"帕帕拉恰"之色，是帕帕拉恰最好的颜色，但是这种颜色比例很罕见，因此也可以选择粉色和橙色的比例为55%和45%的，算是第二优质颜色。

The certificate subdivides Padparadscha into two types: "sunrise" and "sunset". Sunrise refers to the medium to strong saturation, mainly pink (80% pink +20% orange) Padparadscha. Sunset refers to a medium to strong, orange-tinged (80% orange +20% pink) Padparadscha. When pink and orange each account for 50%, it creates an excellent "Padparadscha" color, which is the best color of Padparadscha, but this color ratio is rare, so you can also choose the ratio of pink and orange 55% and 45%, which is the second best color.

小雨老师经验分享
Experience Sharing from Ms. Xiaoyu

证书主要就看GRS证书，颜色那一栏是日落色还是日出色，这样的颜色标明了它的价值。当然也可以看古柏林、SSEF的证书，但是他们不会给颜色做分级，如果能出帕帕拉恰就已经证明了颜色很不错。

We mainly look at the GRS certificate, to see the color column is the sunset color or the sunrise color, which indicates its value.

Of course, you can also look at Gubelin, SSEF certificate, but they do not grade the color, if they mark Padparadscha, it's approving that the color is very good.

Certificates

sunrise证书

sunset证书

(二) 帕帕拉恰的投资收藏建议
Padparadscha Investment and Collection Suggestions

1. 帕帕拉恰涨价的原因
Reasons for the Price Increase in Padparadscha

　　帕帕拉恰的产量大概只有红宝石产量的1%—1.5%，可见其稀有程度，每出产一颗都是"五万分之一的奇迹"。所以在市面上很少能够看到高品质的帕帕拉恰，大多数市面上的帕帕拉恰都是擦边球的货，要么颜色特别粉，看上去像粉色蓝宝石，要么橘色调特别重，偶尔还带有棕色调。真正的帕帕拉恰是粉色和橘色混合在一起的，大克拉的在市面上非常少见，只有在拍卖会才能看见。而帕帕拉恰独特的颜色又非常吸引宝石爱好者，据2023年统计，帕帕拉恰的热度为34.73%，占到彩色宝石第一。可见帕帕拉恰在宝石界的受欢迎程度，所以涨价是必然趋势。查询了近几年的大型珠宝拍卖会数据，可以看出帕帕拉恰价格持续上涨，7年时间甚至翻涨8倍！帕帕拉恰成为香港佳士得2017年宝石拍卖的焦点。

　　The production of Padparadscha is only about 1-1.5% of the production of rubies, which shows its rarity, and each production is a "miracle of one in 50,000." Therefore, it is rare to see high-quality Padparadscha on the market. Most of the Padparadscha on the market is a marginal ball of goods, either the color is particularly pink, looks like pink sapphire, or the orange color is particularly heavy, and occasionally with brown tones. Real Padparadscha is a mix of pink and orange, and large carats are very rare on the market and can only be seen at auction. The unique color of Padparadscha is very attractive to gem lovers. According to 2023 statistics, the heat of Padparadscha is 34.73%, accounting for the first color gem. It can be seen that the popularity of Padparadscha in the gem world, so the price increase is an inevitable trend. Inquiring the data of large jewelry auctions in recent years, it can be seen that the price of Padparadscha continues to rise, and even increases by 8 times in 7 years! Padparadscha was the focus of Christie's 2017 gem sale in Hong Kong.

2. 帕帕拉恰近几年涨价的大约情况
Information on the Price Increases in Padparadscha in Recent Years

克拉 Carat	2.00—3.00克拉 （元人民币/克拉） 2.00—3.00ct（CNY/ct）	5.00克拉 （元人民币/克拉） 5.00ct（CNY/ct）	10.00克拉 （元人民币/克拉） 10.00ct（CNY/ct）	20.00克拉 （元人民币/克拉） 20.00ct（CNY/ct）
2019年	40,000	100,000	200,000	400,000
2020年	50,000	150,000	250,000	450,000
2021年	80,000	180,000	300,000	500,000
2022年	100,000	250,000	400,000	600,000
2023年	130,000	300,000	550,000	800,000
2024年	150,000	350,000	600,000	900,000

3. 帕帕拉恰的拍卖
Padparadscha Auction Record

时间 Time	拍卖行 Auction House	拍品 Item	成交价 Closing Price
2023年5月 May, 2023	苏富比 Sotheby's	12.43克拉海瑞温斯顿橙粉红色帕帕拉恰戒指，斯里兰卡产，未经加热 12.43 Kerahari Winston Orange Pink Padparadscha Ring, made in Sri Lanka, unheated	成交价 508,000法郎 Sale Price CHF 508,000
2023年3月 March, 2023	苏富比纽约 Sotheby's New York	8.13 克拉帕帕拉恰戒指，斯里兰卡产，未经加热 8.13 CraPadparadscha Ring, made in Sri Lanka, unheated	成交价 44,450美元 Sale Price USD 44,450
2021年12月 December, 2021	邦瀚斯纽约 Bonhams New York	4.58 克拉帕帕拉恰戒指，未经加热 4.58 ClaPadparadscha ring, unheated	成交价 12,115.50美元 Sale Price USD 12,115.50
2018年11月 November, 2018	佳得士日内瓦 Christie's Geneva	14.67克拉卡地亚帕帕拉恰戒指，斯里兰卡产，未经加热 14.67 carat Cartier Padparadscha Ring, made in Sri Lanka, unheated	成交价 68,750法郎 Sale Price CHF 68,750

4. 帕帕拉恰的投资案例
Padparadscha's Investment Case

小雨老师经验分享
Experience Sharing from Ms. Xiaoyu

2016年我还在上学，就很关注帕帕拉恰，因为它的颜色很让我着迷。那个时候我去香港参加国际珠宝展，3克拉的帕帕拉恰才7万元人民币，到2023年的时候，价格变成了30万元人民币。我当时还是学生没有太多钱，就让跟我同行的一个做美容生意的姐姐买了一颗5克拉的，好像是25万元人民币，2022年她生意需要钱，让我帮她找一个买家，最后成交价格是95万元人民币。

In 2016, when I was still in school, I paid attention to Padparadscha because its color fascinated me. At that time, I went to Hong Kong to attend the international jewelry exhibition, and the 3-carat Padparadscha was only 70,000 yuan, and by 2023, the price had become 300,000 yuan. At that time, I was still a student and did not have much money, so I asked a sister who was with me in the beauty business to buy a 5-carat one, which seemed to cost 250,000 yuan. In 2022, she needed money for her business and asked me to help her find a buyer, and the final transaction price was 950,000 yuan.

5. 帕帕拉恰的投资建议
Padparadscha's Investment Advice

（1）权威证书
Authoritative Certificate

权威的宝石实验室的证书，保证帕帕拉恰的品质。

The certificate from an authoritative gemological laboratory guarantees the quality of Padparadscha.

（2）高品质帕帕拉恰
High-quality Padparadscha

未经热处理，净度好，粉色和橘色融合在一起的高品质帕帕拉恰。

high-quality padparadscha without heat treatment, good clarity, pink and orange colors.

（3）了解市场
Understand the Market

了解国际市场趋势和价格波动，多看看拍卖价格。在购买前咨询珠宝专家或名贵珠宝投资顾问。

Learn about international market trends and price fluctuations and take a look at auction prices. Consult a jewelry expert or a fine jewelry investment advisor before buying.

帕帕拉恰是蓝宝石中特别稀有的一个品种，它最重要的价值就是来自它的颜色，颜色一定是橘色、粉色都有并能混合在一起。不论是斯里兰卡还是缅甸，还是马达加斯加都有高品质帕帕拉恰。GRS证书会对橘色偏多的叫做日落色，粉色偏多的叫做日出色，每个人各有所爱，这两种颜色在国际市场上都非常受欢迎。像3克拉、5克拉的高品质帕帕拉恰都非常难得，有保值增值的潜力，收藏级的帕帕拉恰肯定是5克拉以上的，目前的市面价格肯定是过百万元人民币的。2018年一颗20克拉的高品质帕帕拉恰，拍卖价格已经达到2000万元人民币。英国一些皇室求婚都会用帕帕拉恰，一些国际珠宝大牌高定珠宝系列也会用到帕帕拉恰，在贵族和富人的圈子里，帕帕拉恰是认可度非常高的名贵珠宝。

Padparadscha is a particularly rare variety of sapphire, its most important value is from its color, the color must be orange, pink and can be mixed together. From Sri Lanka to Myanmar to Madagascar, there are high quality Padparadscha. GRS certificate will name the ones with more orange 'sunset' color, more pink ones called 'sunrise'. Everyone has their own love, and these two colors are both very popular in the international market. Like 3 carat, 5 carat high-quality Padparadscha are very rare, there is the potential to preserve and increase value. Collection grade Padparadscha is definitely more than 5 carat, the current market price is definitely more than one million yuan. In 2018, the auction price of a 20-carat high-quality Padparadscha has reached 20 million yuan. Some British royal marriage proposals will use Padparadscha, some international jewelry brands high-end jewelry series will also use Padparadscha. In the circle of nobility and the rich, Padparadscha is a very high degree of recognition of the luxury jewelry.

Appreciating Paraiba

第五章
05 Chapter Five

帕拉伊巴
Paraiba

一、帕拉伊巴的基本特征
Basic Characteristics of Paraiba

帕拉伊巴的英文名叫Paraiba，是碧玺中的爱马仕，也就是碧玺中最值得收藏的品种，产自巴西和莫桑比克，以其罕见的霓虹蓝色而闻名。1989年，宝石开采者在帕拉伊巴山丘发现了这一独特的碧玺，呈现出明亮的水蓝色，引起宝石界轰动。帕拉伊巴碧玺的特别之处在于其明亮而霓虹般的蓝绿色。它是由铜锰共同致色的电气石，包含蓝色（电光蓝、霓虹蓝、紫蓝色）、蓝绿色、绿蓝色或者绿色色调，人们用"Neon-Blue"或"Paraiba-Blue"来描述这种前所未见的宝石色彩。

Paraiba is the Hermes of tourmalines, and also the most collectible of the tourmalines. It is produced in Brazil and Mozambique and known for its rare neon blue color. In 1989, gem miners discovered this unique tourmaline in the Paraiba hills, showing a bright aquamarine blue, causing a sensation in the gem world. The special feature of Paraiba tourmaline is its bright and neon blue-green color. It is a combination of copper and manganese tourmaline color, including Blue (electric blue, Neon Blue, purple blue), blue-green, green-blue or green tones, people use the term "neon blue" or "Paraiba-Blue" to describe this unprecedented gemstone color.

Basic characteristics of Paraiba

帕拉伊巴裸石

帕拉伊巴碧玺戒指

帕拉伊巴碧玺戒指

帕拉伊巴矿区

帕拉伊巴矿区买卖

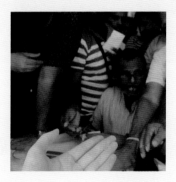
帕拉伊巴买卖

Basic Characteristics of Paraiba

What makes Paraiba tourmaline special is its bright and neon-like blue-green color, shaped by the presence of copper and manganese.

二、如何鉴赏和投资收藏帕拉伊巴
How to Appreciate and Invest in the Paraiba Collection

  我们主要从以下 7 个维度去鉴赏和投资一颗帕拉伊巴，我在每一个鉴赏指标分析里都给出了目前的市场情况以及相应指标的价格影响，也参考了一些拍卖价格，大家可以参考。

  We are going to appreciate and invest a Paraiba from the following seven aspects. In each appreciation index analysis, I have given the current market situation and the price impact of the corresponding indicators, and also referred to some auction prices for your reference.

（一）帕拉伊巴的鉴赏
Appreciation of Paraiba

1. 颜色
Color

帕拉伊巴碧玺的颜色占整颗帕拉伊巴价值的 50%，是其最重要的特征之一，它的迷人之处正是来自那种独特的"霓虹蓝"电光色，且为最贵的颜色。

The color of the Paraiba tourmaline accounts for 50% of the value of the whole Paraiba and is one of its most important features. Its charm is precisely from the unique "neon blue" electric color, and is the most expensive color.

绿色帕拉伊巴
Green Paraiba

帕拉伊巴碧玺的绿色常常带有一定的蓝色色调，呈现出明亮而鲜艳的绿蓝色。

The green of Paraiba tourmaline is often tinged with a certain blue hue, showing a bright and vivid greenish-blue color.

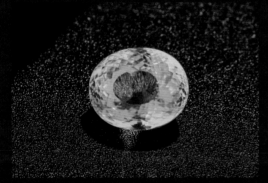

绿色帕拉伊巴

紫色帕拉伊巴
Purple-blue Paraiba

有些帕拉伊巴碧玺可能显示出紫蓝色的色调，这种颜色可能同时包含蓝色和紫色的元素，而且紫色的帕拉伊巴一般是没有经过热优化的。

Some paraiba tourmalines may show a purple-blue hue, which may contain both blue and purple elements, and purple paraiba is generally not thermally optimized.

紫色帕拉伊巴

蓝色帕拉伊巴
Blue Paraiba

帕拉伊巴碧玺的蓝色可以是明亮的水蓝色，有时也会带有一些绿色或紫色的色调。霓虹蓝是帕拉伊巴碧玺最受欢迎和珍贵的颜色之一，呈现出明亮而耀眼的蓝色，通常带有一定的绿色调，同时散发出特殊的电光。

The blue of Paraiba tourmaline can be a bright aqua blue, sometimes with some green or purple tints. Neon blue is one of the most popular and prized colors of Paraiba tourmaline, presenting a bright and dazzling blue, often with a certain green tinge, while emitting a special electric glow.

蓝色帕拉伊巴

霓虹蓝帕拉伊巴
Neon Blue Paraiba

Neon Blue Paraiba

霓虹蓝电光帕拉伊巴

电光色为最佳，与普通的蓝碧玺不同，帕拉伊巴碧玺的颜色明亮而鲜艳，给人一种清澈而耀眼的感觉。帕拉伊巴碧玺可以在微弱的光线下仍然散发出明亮的光芒和独特的火光，这是其独特之处。这种霓虹蓝色彩的帕拉伊巴碧玺非常受人欢迎，也是价值最高的。它的颜色与一般蓝碧玺不同，没有暗沉的感觉，而是鲜明明亮的蓝色，略带绿色。有人形容帕拉伊巴碧玺的颜色像湖水般凝固、清澈而宜人。

The electric color is the best, and unlike ordinary blue tourmaline, the color of Paraiba tourmaline is bright and vivid, giving a clear and dazzling feeling. Paraiba tourmaline can still emit a bright light and a unique fire in the weak light, which is its unique feature. This neon blue Paraiba tourmaline is very popular and is the most valuable. Its color is different from the general blue tourmaline, there is no dark feeling, but bright blue, and slightly green. Some people describe the color of Paraiba tourmaline as solid, clear and pleasant like lake water.

除了颜色，电光也是评估帕拉伊巴碧玺质量的重要标准。拥有电光的帕拉伊巴碧玺价格较高，电光的存在是区分顶级帕拉伊巴和普通帕拉伊巴的关键特征。即使两颗帕拉伊巴碧玺的颜色相同，但有电光的帕拉伊巴的价格可以达到没有电光的三倍甚至更高，因为电光是顶级帕拉伊巴的重要标志之一。

In addition to color, electric light is also an important criterion for evaluating the quality of Paraiba tourmaline. Paraiba tourmalines with electric light have a higher price, and the presence of electric light is a key feature that distinguishes top paraiba from ordinary paraiba. Even if two paraiba tourmaline are the same color, the price of Paraiba with electric light can reach three times or more than that without electric light, because electric light is one of the important signs of top paraiba.

小雨老师经验分享
Experience Sharing from Ms. Xiaoyu

帕拉伊巴最重要的价值就是它的颜色，在其他参数都相同的情况下，巴西产的帕拉伊巴颜色非常浓郁，即便它的净度差一点，价格还是远远高过莫桑比克产的帕拉伊巴。打个比方，1克拉的巴西帕拉伊巴2018年的时候拍卖就可以拍卖到40万元人民币，但是如果是3克拉的莫桑比克，各方面品质还不错的，也不一定能卖到40万元人民币，所以颜色是帕拉伊巴最重要的价值因素。电光蓝的浓郁蓝色调，目前是市面上价格最高的颜色，10克拉的这种高品质帕拉伊巴，价格可能需要300万元人民币左右。

The most important value of Paraiba is its color. In the case that other parameters are the same, the color of Paraiba made in Brazil is very rich, even if its clarity is less, and the price is far higher than that of Paraiba made in Mozambique. For example, the 1-carat Brazilian Paraiba can be auctioned to 400,000 yuan in 2018, but if it is a 3-carat Mozambique, which has good quality in all aspects, it may not be able to sell 400,000 yuan. So the color is the most important value factor of Paraiba. The intense blue hue of electric blue is currently the most expensive color on the market, and the price of this high-quality 10-carat Paraiba may cost about 3 million yuan.

2.净度
Clarity

包裹体的大小、数量和位置：包裹体越大、数量越多，对宝石的净度影响就越大。特别是当包裹体位于台面部位时，对宝石净度的影响更为显著。

Size, quantity, and location of inclusions: The larger and more numerous the inclusions, the greater their impact on the clarity of the gemstone. Particularly, when inclusions are located on the table facet, their impact on the gemstone's clarity becomes more significant.

净度优秀的帕拉伊巴

净度差的帕拉伊巴

1. 包裹体的大小、数量和位置 / Size, Quantity and Location of Inclusions

包裹体越大、数量越多，对宝石的净度影响就越大。特别是当包裹体位于台面部位时，对宝石净度的影响更为显著。

The larger and more numerous the inclusions, the greater the impact on the clarity of the stone. Especially when the inclusion is located in the mesa, the influence on the clarity of the gem is more significant.

2. 包裹体的类型和对比度 / Type and Contrast of Inclusions

包裹体的颜色与宝石的差异也会影响宝石的净度。如果包裹体的颜色与宝石相差较大，对宝石的净度影响会更明显。

The difference between the color of the inclusions and the stone can also affect the clarity of the stone. If the color of the inclusion is different from that of the gem, the impact on the clarity of the gem will be more obvious.

3. 未愈合的裂隙 / Unhealed Cracks

如果宝石内部有未愈合的裂隙，将降低宝石的耐久性，容易受到损伤。

If there are unhealed cracks inside the gem, it will reduce the durability of the gem and make it vulnerable to damage.

 小雨老师经验分享
Experience Sharing from Ms. Xiaoyu

帕拉伊巴的净度很不好，在帕拉伊巴碧玺的评定中，净度并不像其他宝石那样具有明确的定义和划分。有时候帕拉伊巴内部特殊的包裹体反而可以呈现出电光质感，反而是加分项。取而代之的是，人们更关注帕拉伊巴碧玺的颜色的鲜艳度、透明度和电光等特征。对于帕拉伊巴的净度，我经常说，如果在社交距离不影响它整体的美观的话，我们用不着去强调它一定要全干净玻璃体，因为这样的帕拉伊巴真的太少了。如果存在这样的超级干净的帕拉伊巴，那么它的价格将十分的贵。比如一颗5克拉的净度特别好的帕拉伊巴，价格在50万元人民币左右，比它净度差一点点，但是颜色特别好的帕拉伊巴，可能价格反而需要80万元人民币左右。

The clarity of Paraiba is very poor, and in the evaluation of Paraiba tourmaline, clarity is not as clearly defined and divided as other gemstones. Sometimes the special inclusions inside Paraiba can give an electric texture, which is a plus. Instead, people pay more attention to the color of Paraiba tourmaline, such as brightness, transparency and electric light. For the clarity of Paraiba, I often say that if in social distance, there is no effect on its overall beauty, then we do not need to emphasize that it must be completely clean glass body, as such Paraiba is really too few. If such a super-clean Paraiba existed, it would be very expensive. For example, a 5-carat Paraiba with a particularly good clarity, the price is about 500,000 yuan. While if there is a worse in the clarity, but the color is particularly good Paraiba, and the price may come to about 800,000 yuan.

3. 切工
Cut

切工主要表现在对称性和抛光质量等方面。对称性的问题指正侧面轮廓的对称性偏差、台面偏心、底尖偏心、亭部膨胀、刻面畸形和刻面尖点不尖等。抛光质量指的是宝石表面的抛光程度。

The cutting is mainly manifested in symmetry and polishing quality. The symmetry problems include the symmetry deviation of the side profile, the eccentricity of the mesa, the eccentricity of the bottom tip, the bulge of the pavilion, the facet deformity and the non-cusp of the facet. Polishing quality refers to the degree of polishing of the gem's surface.

露底帕拉伊巴

小雨老师经验分享
Experience Sharing from Ms. Xiaoyu

帕拉伊巴在只要没有严重的漏底，市场上都是可以接受的，帕拉伊巴碧玺的选择并不仅仅取决于其形状，而更关注其独特的颜色、火彩和整颗宝石的整体质量。无论哪种形状和琢磨方式，帕拉伊巴碧玺都能散发出令人惊叹的光彩。帕拉伊巴的火彩需要好的切割才能把它衬托出来，因为它内部含有那种电光质感，比如说两颗5克拉的高品质的帕拉伊巴，其他条件都相同的情况下，一颗帕拉伊巴价值50万元人民币，但是切割特别好的话，电光质感就会强一些，那价格可能就需要60万元人民币。

Paraiba tourmaline is acceptable in the market as long as there is no serious leakage, and the selection of Paraiba tourmaline is not only determined by its shape, but more concerned with its unique color, fire color and the overall quality of the whole stone. No matter what shape and treatment, Paraiba tourmaline can emit a stunning brilliance. Paraiba fire color needs a good cutting to bring it out, because it contains the kind of electric texture. For example, two 5 carat high-quality Paraiba, with all other conditions are the same, one paraiba worth 500,000 yuan. But the other one with particularly good cutting, and the electric texture will be stronger, so the price may come to 600,000 yuan.

4. 克拉重量
Carat Weight

由于帕拉伊巴碧玺的开采困难和产量稀少，因此较大克拉数的帕拉伊巴碧玺裸石非常罕见。一般来说，市场上流通的帕拉伊巴碧玺裸石的克拉数通常在 3 克拉至 10 克拉之间，极少超过 20 克拉。迄今为止，市场上出现过的最大帕拉伊巴碧玺裸石为 100 克拉，这是一个非常罕见和令人惊叹的发现。帕拉伊巴碧玺的产量确实非常有限，每月全球的产量只有数十克拉，而且由于其晶体结构脆弱，开采过程中晶体很容易支离破碎，导致找到的帕拉伊巴碧玺原石几乎没有完整的。因此，能够亲眼见到一颗 5 克拉以上的蓝色调帕拉伊巴碧玺是非常难得的了。这些稀有的大克拉数帕拉伊巴碧玺裸石对于珠宝行业和收藏家来说具有非常高的价值和吸引力。

Because of the difficulty of mining Paraiba tourmaline and the scarcity of production, large carats of Paraiba tourmaline are very rare. In general, the number of carats of Paraiba tourmaline in circulation on the market is usually between 3 and 10 carats, and rarely exceeds 20 carats. The largest bare Paraiba tourmaline that has ever appeared on the market is 100 carats, which is a very rare and amazing find. The production of Paraiba tourmaline is indeed very limited, only tens of carats per month worldwide, and because of its fragile crystal structure, the crystals are easily fragmented during mining, resulting in almost no complete Paraiba tourmaline. Therefore, it is very rare to see a blue Paraiba tourmaline of more than 5 carats in person. These rare large carat Paraiba tourmaline bare stones are of very high value and appeal to the jewelry industry and collectors.

小雨老师经验分享
Experience Sharing from Ms. Xiaoyu

帕拉伊巴的克拉重量同样存在溢价的问题，1到3克拉的帕拉伊巴是好找的，大概价位是小几十万元人民币不等，但如果上了5克拉的高品质帕拉伊巴，价格有可能是要过百万元人民币的。帕拉伊巴的晶体可以长得比较大，所以说我们作为投资和收藏来讲的话，可以买一些大克拉的帕拉伊巴。比如说5克拉的帕拉伊巴，价格在100万元人民币左右，但是10克拉的帕拉伊巴，价格可能是500万元人民币左右，因为它的克拉溢价不是按倍数来增长的，可能是几倍增长的。

Paraiba carat weight also has a premium problem, 1 to 3 carat Paraiba is easy to find, the approximate price is hundreds of thousands of yuan. But if it is a 5 carat high-quality Paraiba, the price may be more than one million yuan. Paraiba crystals can grow relatively large, so we can buy some large carats of Paraiba as an investment and collection. For example, the price of 5-carat paraiba is about 1 million yuan, but the price of 10-carat Paraiba may be about 5 million yuan. This is because its carat premium is not increased by multiple, it may be several times larger.

5. 产地
Place of Origin

帕拉伊巴的产地对价格影响非常大,巴西产的帕拉伊巴价格远超过莫桑比克产的价格。

The origin of Paraiba has a strong influence on the price, and the price of Paraiba made in Brazil far exceeds that of Mozambique.

小雨老师经验分享
Experience Sharing from Ms. Xiaoyu

巴西和莫桑比克两个产地的价差特别大,巴西产地的帕拉伊巴蓝色调更浓郁一些,价格可以达到40万元人民币1克拉,2克拉的巴西帕拉伊巴需要100万元人民币左右。莫桑比克的帕拉伊巴蓝色调浅一些,价格也达到了10万元人民币1克拉。从投资收藏来看,巴西的帕拉伊巴在拍卖上非常受欢迎,但是颜色比较浓郁的莫桑比克帕拉伊巴,也慢慢进入拍卖市场。

The price difference between Brazil and Mozambique is particularly large. The Brazilian origin of Paraiba blue color is stronger, the price can reach 400,000 yuan for 1 carat, and 2 carat Brazilian Paraiba costs about 1 million yuan. Mozambique's Paraiba is a lighter shade of blue, and the price also reaches 100,000 yuan per carat. From the perspective of investment collection, Brazil's Paraiba is very popular at auction, but the Mozambique Paraiba, which is relatively rich in color, is also slowly entering the auction market.

1 巴西 Brazil

巴西帕拉伊巴

被认为是帕拉伊巴碧玺的主要产地,尤其是巴西巴伊亚州。1988年,巴西帕拉伊巴州的宝石矿区首次发现了这种宝石,其明亮的水蓝色和高含量的铜和锰成分使其与众不同。然而,巴西产的帕拉伊巴碧玺产量稀少,裸石颗粒小且净度较低,因此市场上的大部分帕拉伊巴碧玺实际上并非来自巴西,而是来自非洲的尼日利亚和莫桑比克。这些产地的帕拉伊巴碧玺可能具有不同的颜色、透明度和光彩效果,因此价格可能会有所差异。

It is considered to be the main source of Paraiba tourmaline, especially in the Brazilian state of Bahia. First discovered in the gemstone mining area of Paraiba state in Brazil in 1988, this gem is distinguished by its bright aqua blue color and high content of copper and manganese components. However, Brazilian Paraiba tourmaline is rare in production, with small bare stone particles and low clarity, so most of the Paraiba tourmaline on the market does not actually come from Brazil, but from Nigeria and Mozambique in Africa. Paraiba tourmalines from these origins may have different colors, transparency and brilliance effects, so prices may vary.

2 莫桑比克 Mozambique

因其特殊的鲜艳蓝绿色而备受珠宝爱好者和收藏家的追捧。这种宝石最早在巴西的帕拉伊巴州发现,因此得名帕拉伊巴。后来,在20世纪90年代初,类似的宝石在莫桑比克的承恩塔区域被发现,被称为莫桑比克帕拉伊巴。莫桑比克帕拉伊巴的蓝色总体要比巴西帕拉伊巴颜色调浅一些。

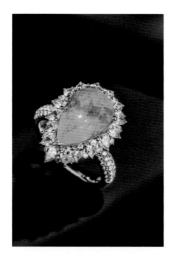

It is highly sought after by jewelry lovers and collectors because of its special vivid blue-green color. This gem was first found in the Brazilian state of Paraiba, hence getting the name Paraiba. Later, in the early 1990s, a similar gem was discovered in the Chengta region of Mozambique, known as Mozambique Paraiba. Paraiba in Mozambique is generally a lighter shade of blue than Paraiba in Brazil.

6.优化
Optimization

帕拉伊巴碧玺的热处理

加热
Heating

一种常见的优化处理方法是加热处理。通过加热处理，可以改善帕拉伊巴碧玺的颜色饱和度和透明度，使其更具吸引力。这种优化处理方法可能会对价格产生积极影响，因为改善后的颜色和外观能够增加宝石的市场价值。帕拉伊巴碧玺通常经过热优化，以改善颜色并消除锰元素引起的紫色。这个低温优化不会留下痕迹，而且颜色十分稳定，使得辨别是否经过处理较为困难。大多数实验室，包括 GIA，通常不会确认碧玺是否经过了热处理。

A common optimization method is heating treatment. By heating treatment, the color saturation and transparency of Paraiba tourmaline can be improved, making it more attractive. This optimized treatment is likely to have a positive impact on prices, as the improved color and appearance increase the market value of the stone. Paraiba tourmaline is usually thermally optimized to improve color and eliminate the purple caused by manganese. This low temperature optimization leaves no marks, and the color is very stable, making it difficult to tell whether it has been treated or not. Most laboratories, including GIA, do not usually confirm whether tourmaline has been heat-treated.

小雨老师经验分享
Experience Sharing from Ms. Xiaoyu

帕拉伊巴的热优化不影响帕拉伊巴的价值，因为国际上默认大多数的帕拉伊巴都是经过热优化的。除非拿去鉴定的人特别说明，帕拉伊巴没有经过任何优化，但是不论这个帕拉伊巴是否优化，都不影响它的整个价值，因为漂亮才是最重要的。比如说一个经过热优化的帕拉伊巴，颜色出现了霓虹蓝的色调，价值远远比一颗没有经过热优化的、但是颜色很普通的蓝色调帕拉伊巴，价值要高很多，所以说帕拉伊巴的热优化不影响它的价格。除非两个都是特别漂亮的，一个是热优化，一个是无热优化的，那么无热优化的价格要高一些。

The thermal optimization of Paraiba does not affect the value of Paraiba, because Paraiba is the international default that most Paraiba are thermal optimized. Unless the person who bring the Paraiba to be evaluated specifically states that the Paraiba has not been optimized in any way. But whether the Paraiba is optimized or not does not affect its overall value, because beauty is the most important thing. For example, a thermally optimized Paraiba has a neon blue color tone, and the value is far higher than a non-thermally optimized Paraiba. But one's color is very ordinary blue tone, and the value is much higher, which refers that the thermal optimization of the Paraiba does not affect its price. Unless the two are particularly beautiful, one is thermal optimization, one is non-thermal optimization, then the price of non-thermal optimization will be higher.

7. 证书
Certificates

不同的证书代表不同的含金量
Different certificates represent different gold content

01 GRS

GRS 对于巴西的帕拉伊巴，会在颜色那一栏，给出蓝到绿色调不等的结论，但如果颜色达到了霓虹色，那么会给出霓虹色，括号中给出"帕拉伊巴色"，或者"帕拉伊巴—霓虹色"，而且在产地那一栏会给出巴西或者莫桑比克。

For Paraiba in Brazil, the GRS will give a conclusion ranging from blue to green in the color column. But if the color reaches neon, then it will give a neon color, giving "Paraiba" in parentheses, or "Paraiba - neon color", and in the origin column will mark Brazil or Mozambique.

02 GUILD

GUILD 所出证书也会给出帕拉伊巴的定级，以蓝色调和绿色调划分为 Blue（蓝色）、Green（绿色）、bluish Green（蓝绿色）、greenish Blue（绿蓝色）、Blue-Green（蓝—绿色）、Green-Blue（绿—蓝色）。对于颜色可达到 Neon（霓虹色）的宝石，GUILD 会在颜色前添加 Neon（霓虹）的描述。

GUILD certificates also give Paraiba a rating. It is divided into Blue, Green, bluish Green, greenish Blue, Blue-Green, Green-Blue. For gems that reach Neon color, GUILD will add before the color with the description of Neon.

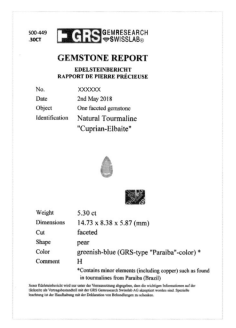

GRS证书

GUILD证书

087

03 SSEF

SSEF证书

SSEF 会依据帕拉伊巴的颜色，使用描述性术语来描述，例如蓝色、浅蓝色、绿蓝色、绿色，但不会使用任何商业术语。

SSEF will use descriptive terms such as blue, light blue, greenish blue, green to describe the color of Paraiba, but will not use any commercial terms.

小雨老师经验分享
Experience Sharing from Ms. Xiaoyu

主要看证书里的2个地方，一个是颜色，能否给出帕拉伊巴—霓虹色，一个是产地是不是巴西。其实我们作为在珠宝行业的内行，一般看到颜色特别浓郁的，我们就知道这个产地大概是在巴西，我们会取出一个很好的证书，比如说像GRS。但如果这个颜色还蛮浅的，那我们可能就出一个吉尔德的证书。产地是莫桑比克的，宝石越贵，出的证书的级别就越高，基本上是这个逻辑。当然了，还是那句话，一张证书不能完全证明这个宝石的价值，主要还是要看这个宝石具体的品相是怎么样的。

　　We mainly look at the two places in the certificate, one is the color, whether it can be marked with Paraiba - neon color, one is whether the origin is Brazil. In fact, as experts in the jewelry industry, when see the color is particularly strong, we generally know that the origin is probably in Brazil. We will go for a good certificate, such as GRS. But if it's a light color, we might just issue a GUILD certificate. If the origin is Mozambique, the gem will be more expensive, and the higher the level of the certificate. This basically is the logic. Of course, the same old saying, a certificate can not completely prove the value of this gem, it mainly depends on how is the specific appearance of this gem.

(二)帕拉伊巴的投资收藏建议
Paraiba Investment and Collection Suggestions

1.帕拉伊巴的涨价原因
Reasons for Paraiba's Price Increase

高品质的帕拉伊巴因为疯狂挖掘几近绝矿，因为帕拉伊巴特殊的晶体结构，它的裂隙发育比较明显，在开采的时候又不能通过爆破的方式获得帕拉伊巴，只能手工开采，所以效率极低。多数开采出的帕拉伊巴都是碎片状的原石，超过5克拉的品质完好的帕拉伊巴已属于精品。拍卖场上也很难看到帕拉伊巴的身影，比红蓝宝、祖母绿还要稀缺。有时候我出去讲课，讲到帕拉伊巴，很多投资人说都没看到过，我就说我们职业宝石鉴定师，在拍卖会上也很难看到几颗高品质的帕拉伊巴。从1989年帕拉伊巴被发现开始，在美国图桑展上展示后的一周时间内从200美元涨到2000美元，到2024年更是几十倍的涨价。对于名贵珠宝而言，高品质霓虹色帕拉伊巴、霓虹色尖晶石涨幅算是相对小的了，他们的颜色表现力太好了，其他宝石无法替代，因为小众才更有投资潜力。

Because of crazy mining, high-quality Paraiba is almost exhausted. Because of the special crystal structure of Paraiba, its crack development is more obvious, and can not be obtained by blasting at the time of mining Paraiba. It can only be mined manually, so the efficiency is extremely low. Most of the Paraiba that is mined is rough in fragments, so a Paraiba weighing more than 5 carats and in good condition is considered excellent. It is also difficult to see Paraiba on the auction house, which is more scarce than rubies, sapphires and emeralds. Sometimes when I go out to give a lecture and talk about Paraiba, many investors say they have not seen it. And I told them that even for us professional gemstone appraisers, it is difficult to see a few high-quality Paraiba at auction. Since Paraiba was discovered in 1989, it has risen from $200 to $2,000 within a week after it was displayed at the Tusan Exhibition in the United States, and it has increased by dozens of times by 2024. For luxury jewelry, price increase rate for high-quality neon color Paraiba, neon color spinel are relatively small. Because their color expression is too good, and other gems can not be replaced. Because of the minority, it has more investment potential.

2. 帕拉伊巴近几年大约价格涨幅
Approximate Price Increase in Paraiba in Recent Years

重量（克拉） Weight(Carat)	2.00-3.00克拉 （元人民币/克拉） 2.00-3.00ct（CNY/ct）	5.00克拉 （元人民币/克拉） 5.00ct（CNY/ct）	10.00克拉 （元人民币/克拉） 10.00ct（CNY/ct）	20.00克拉 （元人民币/克拉） 20.00ct（CNY/ct）
2019 年	30,000	60,000	150,000	200,000
2020 年	45,000	80,000	180,000	300,000
2021 年	50,000	100,000	250,000	400,000
2022 年	60,000	130,000	300,000	480,000
2023 年	80,000	150,000	350,000	550,000
2024 年	42,000	180,000	450,000	600,000

3. 帕拉伊巴投资案例
Paraiba Investment Case

2018年我还在上学的时候，有一个做教育培训的老板跟我说想配置名贵珠宝，他很喜欢蓝色系的宝石，之前也投资过蓝宝石，我就推荐了帕拉伊巴。他说他在国内都没有见过这种宝石，2015年在美国见过，价格很贵。我说帕拉伊巴现在的矿产量还在不断下降，以后会更贵，然后他就花100万元左右配置了一颗高品质的帕拉伊巴。到2022年的时候他在北京卖出了很高的价格，因为2022年帕拉伊巴在国内已经火起来了，很多宝石爱好者都在收藏帕拉伊巴了。

In 2018, when I was still in school, a boss who did education training told me that he wanted to configure luxury jewelry. He liked blue stones very much, and had invested in sapphire before. I recommended Paraiba. He said he had not seen the stone in China, but had seen it in the United States in 2015, and it was expensive. I said that Paraiba's current mine production is still declining, and it will be more expensive in the future, and then he spent about 1 million yuan to buy a high-quality Paraiba. By 2022, he sold at a high price in Beijing, because in 2022, Paraiba has become popular in China, and many gem lovers are collecting Paraiba.

4. 帕拉伊巴的拍卖
Paraiba Auction

时间 Time	拍卖行 Auction House	拍品 Item	成交价 Closing Price
2023年5月 May, 2023	佳士得日内瓦 Christie's Geneva	9.27 克拉莫桑比克帕拉伊巴碧玺戒指，低温加热处理 9.27 carat Mozambique Paraiba tourmaline ring, cryogenic heating treatment	成交价 352,800法郎 Sale Price CHF 352,800
2023年2月 February, 2023	苏富比香港 Sotheby's Hong Kong	11.45克拉莫桑比克帕拉伊巴碧玺戒指，低温加热处理 11.45 carat Mozambique Paraiba tourmaline ring, low temperature treatment	成交价 1,397,000港元 Sale Price HKD 1,397,000
2022年5月 May, 2022	佳士得日内瓦 Christie's Geneva	9.04 克拉枕形帕拉伊巴碧玺戒指，巴西，低温加热处理 28.79 carat Paraiba tourmaline, bright oval modified cut	成交价 32,760法郎 Sale Price CHF 32,760
2018年5月 May, 2018	佳士得日内瓦 Christie's Geneva	9.04 克拉枕形帕拉伊巴碧玺戒指，巴西，低温加热处理 9.04 carat pillow Paraiba tourmaline Ring, Brazil, low temperature treatment	成交价 396,500法郎 Sale Price CHF 396,500

5.帕拉伊巴的投资收藏建议
Paraiba's Investment Collection Advice

(1) 权威证书
Certificate of Authority

您购买到真正的帕拉伊巴碧玺，最好有可靠的证书，如GRS、SSEF、AIGS、古柏林等权威机构的国际证书。

When you buy a genuine Paraiba tourmaline, it is best to have a reliable certificate, such as the international certificate of GRS, SSEF, AIGS, Gubelin and other authorities.

(2) 高品质帕拉伊巴
High-quality Paraiba

蓝绿色和高透明度的帕拉伊巴碧玺在市场上更受欢迎，具有更高的价值。最好选择产地是巴西或莫桑比克的。

The blue-green and high transparency Paraiba tourmaline is more popular in the market and has a higher value. The best choice is made in Brazil or Mozambique.

(2) 了解市场
Understand the Market

定期关注市场上的帕拉伊巴碧玺价格和需求变化，主要可以通过佳士得、苏富比的拍卖，还有大型国际珠宝展，了解其市场趋势和评估情况。这将有助于您做出更明智的收藏和投资决策。也可以咨询名贵珠宝投资顾问。

Keep a regular eye on the market for Paraiba tourmaline prices and changes in demand, mainly through Christie's, Sotheby's auctions, as well as major international jewelry shows, to understand its market trends and assessments. This will help you make more informed collection and investment decisions. You can also consult a fine jewelry investment adviser.

小雨老师特别提醒
Special Reminder from Ms. Xiaoyu

帕拉伊巴其实在欧美市场非常地受欢迎,您可以看到像尚美这样的大品牌,他们的高定珠宝系列都会用到帕拉伊巴碧玺。霓虹蓝色调的、有电光质感的颜色是最值得收藏的。目前像那样的5克拉到10克拉的电光质感的霓虹蓝,在市面上一颗难求,非常受欢迎,如果是莫桑比克的10克拉,价格在300万元左右,但如果是巴西的10克拉的帕拉伊巴,价格过千万元。

Paraiba is actually very popular in the European and American markets. You can see big brands like Chammet, their use Paraiba tourmaline in their high-end jewelry collection. Neon blue tones with an electric glow are the best colors to collect. At present, the neon blue of 5 to 10 carats of electric light texture like that is difficult to find in the market and very popular, if it is a 10 carats of Mozambique, the price is about 3 million yuan, but if it is a 10 carats of Brazil's Paraiba, the price is more than 10 million yuan.

收藏帕拉伊巴的时候,最重要的就是看颜色,不论它是巴西的产地还是莫桑比克的产地。如果是巴西的,即使重量比较小,比如只有3克拉,但是颜色很浓郁,也是非常值得收藏的。但如果是莫桑比克的,我们为了保值增值,增大我们的收益,可以考虑买一个5克拉以上的、颜色浓郁一点的,那也是比较有收藏价值的,还是要看具体的宝石的品相。

When collecting Paraiba, the most important thing is to look at the color, whether it is made in Brazil or Mozambique. If it is Brazilian, even if the weight is relatively small, such as only 3 carats, but the color is very rich, it is also very worth collecting. However, if it is Mozambican, in order to preserve and increase the value of our income, we can consider buying a color of more than 5 carats, which is more collectible value, or to see the specific gems appearance.

The Depth of Emerald

第六章
06 Chapter Six

祖母绿
Emerald

一、祖母绿的基本特征
Basic Characteristics of Emerald

祖母绿的英文是"Emerald"。高品质的祖母绿通常具有深绿色、高透明度和良好的光线效果。其中，铬和（或）钒是导致翠绿色的主要元素。祖母绿的莫氏硬度为7.5，较为坚硬，但也具有一定的脆性。祖母绿具有玻璃光泽，通常透明至半透明。祖母绿主要的产地包括哥伦比亚、巴西、赞比亚和津巴布韦等。哥伦比亚的木佐（Muzo）和契沃尔（Chivor）地区被认为是最受欢迎的祖母绿产地之一。祖母绿作为一种珍贵的宝石，具有较高的投资和收藏价值。祖母绿在欧美国家非常受皇室的欢迎，也是拍卖场和国际珠宝大牌的宠儿。很多人觉得祖母绿跟绿色的翡翠很像，其实仔细观察，他们从颜色上也能区分，高品质的祖母绿是深绿色带有一点黄色调和蓝色调，而高品质的翡翠是正阳绿色，如果颜色过深，反而价值变低。从火彩上看，祖母绿和翡翠也不一样，高品质的祖母绿是玻璃体，闪闪发亮。而翡翠是玻璃种，并不会闪闪发亮。

High-quality emeralds usually have a dark green color, high transparency, and good light effects. Among them, chromium and/or vanadium are the main elements that cause emerald green. Emeralds have a Mohs hardness of 7.5, which is relatively hard, but also has a certain brittleness. Emeralds have a glassy luster and are usually transparent to translucent. The main producers of emeralds include Colombia, Brazil, Zambia and Zimbabwe. The Muzo and Chivor regions of Colombia are considered to be among the most popular sources of emeralds. As a precious gemstone, emerald has high investment and collection value. Emeralds are very popular with the royal family in Europe and the United States, and are also the beloved one of auction houses and international jewelry brands. Many people feel that emeralds are very similar to green jade, in fact, if you carefully observe, they can also distinguish from the color. High-quality emeralds are dark green with a little yellow and blue tone, and high-quality emeralds are bright green, if the color is too deep, the value becomes low. From the fire color, emeralds and jade are not the same, high-quality emeralds are vitreous, shiny. Jadeite is a kind of glass and does not shine.

Basic Characteristics of Emerald

赞比亚祖母绿耳环

小雨老师矿区宝石交易

小雨老师矿区宝石交易

无油祖母绿戒指

祖母绿耳钉

小雨老师矿区宝石交易

Basic Characteristics of Emerald

The color of high-quality emerald should be a rich green, reaching at least VVG for investment value.

Emerald

二、如何鉴赏和投资收藏祖母绿
How to Appreciate and Invest in the Emerald Collection

　　我们主要从以下7个维度去鉴赏和投资祖母绿，我在每一个鉴赏指标分析里都给出了目前的市场情况以及相应指标的价格影响，也参考了一些拍卖价格，大家可以参考。

　　We mainly appreciate and invest in emeralds from the following seven aspects. In each appreciation index analysis, I have given the current market situation and the price impact of the corresponding indicators, and also referred to some auction prices for your reference.

（一）祖母绿的鉴赏
Appreciation of Emerald

1.颜色
Color

是指宝石有哪些颜色。
It refers to the color of the stone.

绿色
Green
高品质的祖母绿的颜色是浓郁的绿色调，颜色达到VVG才能有保值增值的价值，市面上大多是浅绿色，只能起到一定的佩戴作用，不能保值。祖母绿的颜色根据产区的不同，会有一些细微的变化，哥伦比亚的祖母绿往往会带有一点点黄色调，而赞比亚的祖母绿往往会带有点蓝色调。

Green
The color of high-quality emerald is a strong green tone, the color reaches VVG to have the value of preservation and appreciation. Most of them in the market are light green, can only play a certain role in wearing, which can not maintain value. The color of emeralds varies slightly from region to region, with Colombian emeralds often having a slight yellow tinge, while Zambian emeralds tend to have a slight blue tinge.

黄色调哥伦比亚祖母绿

蓝色调赞比亚祖母绿

**小雨老师
经验分享**
Experience Sharing
from Ms. Xiaoyu

祖母绿的颜色占到它整个价值的50%，值得收藏的祖母绿颜色一定要达到VVG以上。VVG的意思就是非常深绿的绿色，在GRS的证书上都会体现这个颜色的评级，如果颜色很浅的话，就是GREEN，即绿色的意思。

The color of the emerald accounts for 50% of its entire value, and the emerald color worth collecting must reach above VVG. VVG means very dark GREEN. In the GRS certificate, it will reflect this color rating, if the color is very light, it will be marked GREEN, that is, the meaning of green.

祖母绿
E merald

2.净度
Clarity

是指宝石的内外的纯净程度,是否含有包裹体。
It refers to the degree of purity of the stone inside and outside, whether it contains inclusions.

1 包裹体的大小、数量和位置
Size, Quantity and Location of Inclusions

包裹体越大、数量越多,对宝石的净度影响就越大。特别是当包裹体位于台面部位时,对宝石净度的影响更为显著。

The larger and more inclusions, the greater the impact on the clarity of the stone. Especially when the inclusion is located in the mesa, the influence on the clarity of the gem is more significant.

2 包裹体的类型和对比度
Type and Contrast of Inclusions

包裹体的颜色与宝石本身的颜色很相近的话,对宝石的净度影响不大。如果包裹体的折射率和颜色与宝石相差较大,对宝石的净度影响会更明显。

If the color of the inclusion is very close to the color of the stone itself, it has little effect on the clarity of the stone. If the refractive index and color of the inclusion are different from that of the gem, the impact on the clarity of the gem will be more obvious.

3 未愈合的裂缝
An Open Fissure

如果宝石内部有未愈合的裂隙,将降低宝石的耐久性,容易受到损伤。

If there are unhealed cracks inside the stone, the durability of the stone will be reduced and it will be vulnerable to damage.

4 丝状包裹体
Filamentous Inclusions

少量丝状包裹体存在于宝石内部不会影响宝石的净度质量,反而可以改善宝石的外观。

The presence of a small amount of filamentous inclusions inside the stone does not affect the clarity quality of the stone, but can improve the appearance of the stone.

祖母绿本来天生就裂隙发育很明显,所以祖母绿有个别称叫做秘密花园。像赞比亚的祖母绿通常会含有黑色的包裹体,而哥伦比亚的祖母绿会含有长针状或者管状的包裹体,其实这些都没有关系,这些都是天然宝石的特征。在选择投资收藏祖母绿的时候,应尽量选择黑点不在台面上而是在底部或者背部、针状包裹体或者小裂隙也在底部或者背部的,尽量保持祖母绿的台面比较干净,这样的祖母绿就比较好卖。但是没法过分苛求祖母绿完全干净,因为几乎做不到。

Emeralds are naturally crevular development, which is very obvious, so emeralds have a nickname called secret garden. Emeralds from Zambia usually have black inclusions, while emeralds from Colombia have long needle or tube inclusions, but it doesn't matter, these are the characteristics of natural gemstones. When choosing to invest in the collection of emeralds, you should try to choose those black spots not on the mesa but at the bottom or back, needle inclusions or small cracks are also at the bottom or back, and try to keep the emeralds' countertops relatively clean, so that emeralds are better to sell. But you can't expect emeralds to be completely clean, because they almost never are.

净度优秀的祖母绿

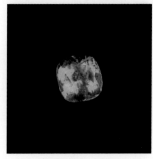

净度较差的祖母绿

小雨老师经验分享
Experience Sharing from Ms. Xiaoyu

净度对祖母绿的价格影响还是蛮大的，因为祖母绿天生裂隙发育比较明显，只要一颗祖母绿相对干净一点，它的价格就贵很多。打个比方，一颗3克拉的特别干净的祖母绿，目前的价格可能要60万元人民币，但如果它有一些肉眼还是能看见的裂隙，那么它的价格大概是30万元人民币。

The impact of clarity on the price of emeralds is still quite large, because emeralds natural crack development is more obvious. As long as an emerald is relatively clean, its price is a lot more expensive. For example, a 3-carat emerald that is particularly clean may cost 600,000 yuan today, but if it has some cracks that are still visible to the naked eye, it will cost about 300,000 yuan.

3.切工
Cut

指宝石外表切割的对称度和工整度。
Refers to the symmetry and fineness of the cut of the gem.

祖母绿有自己专属切割方式，就是一个小八角儿的形状，叫祖母绿切割。祖母绿其实还有其他切割，比如圆形、枕形，但是总体来说，经典的祖母绿切割要比圆形的祖母绿或者枕形的祖母绿价格贵很多。还有一个祖母绿的切割叫素面切割，这样的祖母绿一般都是净度不好的，它就会做成素面有一个弧度，如果净度特别好的，它就会做成刻面，祖母绿切割就是一个刻面切割。刻面切割很容易反映出祖母绿内部的瑕疵。

Emeralds have their own exclusive cutting method, that is, a small octagonal shape, called emerald cutting. Emeralds actually have other cuts, such as round, pillow shape, but in general, the classic emerald cut is much more expensive than the round emerald or pillow shape of the emerald. There is also an emerald cut called plain cutting. Such emeralds are generally not clear, it will be made into a plain face with an arc, if the clarity is particularly good, it will be made into faceted. Emerald cutting is a faceted cutting, which can easily reflect internal flaws in the emerald.

切工优秀的祖母绿

切工较差的祖母绿

小雨老师经验分享
Experience Sharing from Ms. Xiaoyu

如果是从投资收藏的角度，我个人建议尽量选择祖母绿切割，因为这个切割是最能反映祖母绿的净度的，非常地经典，在二次交易的时候也会更好流通。通常情况下，在其他条件都差不多的情况下，一颗2克拉的祖母绿切割的可能要20万元人民币左右，但是如果是枕形切割或者圆形切割的，可能只需要14万元人民币左右。

If it is from the perspective of investment collection, I personally recommend that you try to choose emerald cutting, because this cutting is the most reflective of the clarity of the emerald, very classic, and it will be better circulated in the second transaction. Under normal circumstances, in the case of other conditions are similar, a 2-carat emerald cutting may be about 200,000 yuan, but if it is pillow cutting or circular cutting, it may only cost about 140,000 yuan.

4. 克拉重量
Carat Weight

祖母绿是本身就可以长很大的晶体，所以我们在收藏的时候尽量选择克拉数量较大的祖母绿。我们首先是能保证颜色是VVG或者木佐，在颜色恒定的情况下，尽量选择大颗粒的晶体好的祖母绿，祖母绿在3克拉以上，我都觉得还是蛮有保值的能力。

Emeralds are crystals that can grow large on their own, so we try to choose emeralds with a larger number of carats when we collect them. We can first ensure that the color is such as VVG color or the color of the Musa, then in the case of constant color, try to choose large particles of crystal good emerald. For emerald in 3 carats or more, I think it obtains the ability to preserve value.

祖母绿戒指

小雨老师经验分享
Experience Sharing from Ms. Xiaoyu

我们在看以往的拍卖纪录的时候，大多数祖母绿的拍卖一般5克拉、10克拉、20克拉的，这样的祖母绿是非常受拍卖市场欢迎的，当然价格往往也非常高。在红、蓝、绿这三种宝石当中，祖母绿的单克拉价格算是比较高的。

When we look at the previous auction records, most emerald auctions are generally 5 carats, 10 carats, 20 carats, such emeralds are very popular in the auction market, and of course, the price is often very high. Among the three kinds of gems, red, blue and green, the single carat price of emerald is relatively high.

5. 产地
Place of Origin

是指祖母绿的出产地，对价格影响挺大的。
It refers to the origin of the emerald, which has a great impact on the price.

哥伦比亚
Colombia

哥伦比亚的祖母绿目前是在投资拍卖场上最受欢迎的。高品质的哥伦比亚祖母绿的特点就是颜色很深绿，带有一点点的黄色调。而且净度非常好，一般不会有黑点，火彩也非常亮，这是很多藏家朋友最喜欢的亮点。目前哥伦比亚出产祖母绿最出名的是木佐矿区和契沃尔矿区，这两个矿区产出的祖母绿都有非常高的品质。

哥伦比亚
Colombia

哥伦比亚祖母绿

Colombian emeralds are currently the most popular in investment auctions. High quality Colombian emeralds are characterized by a very dark green color with a slight hint of yellow. And the clarity is very good, generally there will be no black spots, the fire color is also very bright, which is the favorite highlight of many collectors friends. At present, the most famous emeralds produced in Colombia are the Muzo mining area and the Chivor mining area, and the emeralds produced in these two mining areas are of very high quality.

② 赞比亚 Zambia

目前赞比亚矿区也可以产出非常高品质的祖母绿,它的黑点非常少,几乎没有,带有一点点蓝色调,火彩也非常亮,市面上会叫它电光蓝的祖母绿,颜色也非常深邃,目前越来越受消费者的欢迎。但是相比于哥伦比亚来讲,产量要大一些,所以从收藏的角度来讲还是不如哥伦比亚的祖母绿。但是赞比亚的保值能力还是很好的,一直在涨价,高品质的赞比亚祖母绿值得大家关注。

At present, Zambia mining area can also produce very high quality emeralds. Its black spots are very few, almost none, with a little blue tone, the fire color is also very bright. In the market, people call it electric blue emeralds, the color is also very deep, currently more and more popular with consumers. However, compared with Colombia, the output is larger, so from the point of view of collection, it is not as good as Colombia's emeralds. But Zambia's ability to maintain value is still very good, has been rising in price, so high-quality Zambian emeralds deserve everyone's attention.

赞比亚祖母绿戒指

③ 巴西 Brazil

巴西的祖母绿在市面上很少提及,因为大多数人对巴西祖母绿的概念就是颜色有点儿发灰,其实特别高品质的巴西祖母绿颜色是非常深邃的,有点儿像哥伦比亚的祖母绿。巴西祖母绿的价格目前还不高,我自己是愿意购买高品质的巴西祖母绿的,我看好未来的发展空间。

Brazilian emeralds are rarely mentioned in the market, because most people's concept of Brazilian emeralds is a little gray color. But in fact, especially high-quality Brazilian emeralds are very deep color, a little like Colombian emeralds. The price of Brazilian emeralds is not high at present, I am willing to buy high-quality Brazilian emeralds, I am optimistic about the

巴西祖母绿戒指

小雨老师经验分享
Experience Sharing from Ms. Xiaoyu

祖母绿的产地对价格的影响还是非常大的，目前在拍卖会上最受欢迎的就是哥伦比亚的高品质祖母绿，其次是赞比亚的祖母绿，打个比方，一颗10克拉的哥伦比亚祖母绿可能要中几百万元人民币。但是一颗十10克拉的赞比亚祖母绿可能只需要几十万元人民币，所以说产地对它们的价值影响非常大。

The origin of the emerald has a very large impact on the price, and the most popular emerald in the auction is Colombia's high-quality emerald, followed by Zambia's emerald, for example, a 10-carat Colombian emerald may be several million yuan. But a ten-carat Zambian emerald may only cost hundreds of thousands of yuan, so the origin of their value is very important.

6.优化

祖母绿的优化就是浸油，它的油量分4种情况，无油、极微油、微油、中油，证书上都会体现油量的多少，油量越少价格越高。祖母绿的油量也能反映这颗祖母绿的净度，比如说这颗祖母绿是无油的祖母绿，证明它几乎没有开放性的裂隙，油进不去；如果这颗祖母绿是中油的祖母绿，证明它开放性的裂隙比较多，油才会进入。

Emerald optimization is immersed in oil, its oil content is divided into four cases, no oil, insignificant oil, minor oil, moderate oil. The certificate will reflect the amount of oil, the less oil the higher the price. The amount of oil in the emerald can also reflect the clarity of this emerald, for example, one emerald is oil-free emerald, proving that it has almost no open cracks, oil cannot enter; If this emerald is an emerald of medium oil, it proves that it has more open cracks, and oil will enter.

无油
No Oil

证书上会体现No oil这样的字样，证明这颗祖母绿是没有浸油的，在其他条件都相同的情况下，它比微油的价格高。

The certificate will show the words "No oil", proof that the emerald is not immersed in oil, with other things being equal, it is higher than the price of oiled.

极微油
Very Insignificant Oil

证书上会体现Insignificant这样的字样。

The certificate will mark the word "Insignificant".

微油
Light Oil

证书上会体现Minor这样的字样。

The certificate will mark the word "Minor".

中油
Moderate Oil

证书上会体现Moderate这样的字样。

The certificate will mark the word "Moderate".

小雨老师经验分享
Experience Sharing from Ms. Xiaoyu

　　我们在选择投资收藏一颗高品质祖母绿的时候，至少要达到微油以上，也就是它的证书会体现微油。油量达到中油或者重油其实都已经不值得我们投资者去收藏了。一颗无油跟一颗微油的祖母绿，在其他条件都相同的情况下，它们的价格差距是非常大的，打个比方，一颗无油的10克拉的高品质哥伦比亚祖母绿要价过千万元人民币，但是一颗微油的哥伦比亚祖母绿只需要小几百万元人民币。

　　When we choose to invest in a high-quality emerald collection, we must at least reach above the minor oil, that is, its certificate will reflect the minor oil. The amount of oil reached moderate oil or heavy oil is actually not worth our investors to collect. An oil-free emerald and a slightly oily emerald, in other conditions are the same, their price gap is very large. For example, an oil-free 10 carat high-quality Colombian emerald is asking for more than 10 million yuan, but a slightly oily Colombian emerald only cost a few million yuan.

GRS无油祖母绿证书

GRS级微油祖母绿证书

GRS微油祖母绿证书

GRS微油至中油祖母绿证书

7.证书
Certificate

不同的证书代表不同的含金量。
Different Certificates Represent Different Gold Content.

01 GRS

GRS证书会体现祖母绿的颜色级别,如果能达到VVG的颜色,会标有VVG的字样,如果只是普通绿色,只会标有green字样。在油量方面,如果这颗祖母绿是无油的,它会标明NO oil。最后就是产地,也会标明是哥伦比亚的或者赞比亚的。

The GRS certificate will reflect the color level of the emerald. If it can reach the color of VVG, it will be marked with the words VVG. If it is only ordinary green, only the words Green will be marked. In terms of oil content, if the emerald is oil-free, it will be marked NO oil. Finally, the country of origin will be indicated as Colombian or Zambian.

GRS祖母绿证书

02 CD

CD证书来自哥伦比亚,在证书上会体现祖母绿的颜色、产地,还有它的油量。但是在国内的市场上,CD证书的推广并没有那么好,大家会更认可GRS证书。但是在哥伦比亚,CD证书是比较权威的证书。

The CD certificate is from Colombia and will show the color of the emerald, its origin, and its oil content. However, in the domestic market, the promotion of CD certificates is not so good, and we will recognize GRS certificates more. But in Colombia, the CD certificate is the more authoritative certificate.

CD祖母绿证书

03
Gubelin

古柏林是在国际拍卖场上非常受欢迎的一个证书，也是非常权威的一个证书。古柏林证书很少会对颜色做VVG的评级，他只会写一个绿色，但是会对这颗祖母绿的整体品质做一个评分，评分越高说明这颗祖母绿的品质越好。

Gubelin is a very popular certificate in international auctions, and a very authoritative certificate. Gubelin certificate rarely does VVG rating for color, they will only write a Green, but will give a score for the overall quality of the emerald. The higher the score indicates that the quality of the emerald is better.

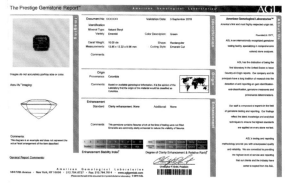

GUBELIN祖母绿证书

04
AGL

AGL非常有权威性，尤其是在欧美国家。它也不会在颜色那一栏具体写一个VVG的这样的评级，但是对颜色会有一个评分，比如说颜色达到了八分、九分，也证明这一颗祖母绿颜色是非常好的。

AGL is very authoritative, especially in Europe and the United States. It will not specifically write a VVG rating in the color column, but there will be a score for the color, for example, the color reached eight or nine points, which also proves that this emerald color is very good.

AGL祖母绿证书

小雨老师经验分享
Experience Sharing from Ms. Xiaoyu

祖母绿的证书差距还是蛮大的，打个比方，如果GRS的证书说这颗祖母绿是无油的，但是把这颗祖母绿拿到古柏林，可能古柏林会出的证书就是有极微油的。所以油量的差距在各个证书之间区别蛮大的。古柏林的证书无油比GRS的证书无油要权威很多，从价格上来讲，一颗古柏林无油的高品质的哥伦比亚祖母绿比GRS无油的哥伦比亚祖母绿要贵上好几倍。

The emerald certificate gap is still quite large, for example, if the GRS certificate says that the emerald is oil-free, but Gubelin may bring out a certificate that writes it is very slight oil. So the difference in the amount of oil between the various certificates is quite different. Gubelin's certificate oil free is much more authoritative than GRS's certificate oil free, and in terms of price, a Gubelin oil-free high-quality Colombian emerald is several times more expensive than GRS oil-free Colombian emerald.

（二）祖母绿的投资收藏建议
Emerald Investment and Collection Suggestions

1.祖母绿涨价的原因
Reasons for the Price Increase of Emeralds

　　Gemfields宝石矿业公司，这家拥有全球70%红宝石资源及25%祖母绿资源的公司及其创立的原石拍卖系统，很大程度上影响着全球彩色宝石的供应链格局及价格走势。2023年5月15日至6月1日，Gemfields举行了2023年第二轮祖母绿原石拍卖，共售出35组拍品，总重26.4万克拉，成交率达100%，总成交额高达4330万美元，打破Gemfields Kagem矿区祖母绿原石拍卖纪录，平均单克拉价格155.9美元，亦为历史最高。给大家看一些祖母绿矿业公司原矿的涨幅情况，大家就明白为什么祖母绿现在价格这么贵了。

　　Gemfields, the company that owns 70% of the world's ruby resources and 25% of the world's emerald resources. And its rough stone auction system, has a large impact on the global supply chain pattern and price trends of colored gems. From May 15 to June 1, 2023, Gemfields held the second round of the 2023 emerald auction, selling a total of 35 lots, a total weight of 264,000 carats, a turnover rate of 100%, a total turnover of 43.3 million US dollars, breaking the Gemfields Kagem mine emerald auction record. The average single-carat price of $155.9 is also an all-time high. Here to show you some of the increase in the emerald mining company's raw ore, you will understand why the price of emeralds is so expensive now.

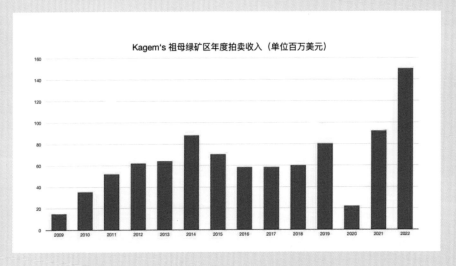

Gemfields2020年—2022年拍卖会
2020—2023 Auction of Gemfields

拍卖会时间 Auction time	2020年11—12月拍卖会 November-December 2020 auction	2021年3—4月拍卖会 March-April 2021 auction	2022年11月拍卖会 November 2022auction
成交重量 Carats Traded	18万克拉	27.1万克拉	27.8万克拉
拍卖总金额 Auction Gross	1090万美元	3140万美元	4330万美元

2.高品质哥伦比亚祖母绿近几年的大约价格涨幅
Approximate Price Increase of High-quality Colombian Emeralds in Recent Years

克拉 Carat	2.00—3.00克拉 (元人民币/克拉) 2.00—3.00ct（CNY/ct）	5.00克拉 (元人民币/克拉) 5.00ct（CNY/ct）	10.00克拉 (元人民币/克拉) 10.00ct（CNY/ct）	20.00克拉 (元人民币/克拉) 20.00ct（CNY/ct）
2019年	25,000	70,000	130,000	200,000
2020年	30,000	80,000	180,000	300,000
2021年	50,000	100,000	250,000	400,000
2022年	60,000	150,000	300,000	500,000
2023年	80,000	200,000	400,000	600,000
2024年	100,000	250,000	500,000	700,000

3.祖母绿的投资案例
Emerald Investment Case

　　我记得特别清楚的一件事是，2019年我还在读博士，有一个跟我认识好几年的姐姐，她之前是做房地产投资的，赚了一些钱，她觉得房地产行业未来会下跌，就问我有什么宝石好投资的，我就给她推荐了5克拉以上的哥伦比亚祖母绿，那个时候价格才30多万元人民币。我看了近10年哥伦比亚祖母绿产区的原矿研报，祖母绿每年的矿产量都在下降，做祖母绿原矿的公司每年价格都在稳步上升。原矿有品质高的，也有品质低的，但是如果平均每个都在涨价，那么挑出来的精品祖母绿未来几年会呈倍数涨价。果然在2023年，5克拉的高品质祖母绿价格需要120万元人民币了，她赚到了钱，请我在北京吃了大餐，哈哈哈。

　　One thing I remember particularly clearly is that in 2019, when I was still a doctoral student, I had a sister whom I had known for several years. She used to do real estate investment and made some money. She felt that the real estate industry would decline in the future, so she asked me what gems she could invest in. I recommended her 5 carat above Colombian emerald. At that time, the price was just over 300,000 yuan. I have read the research report of the raw ore in the emerald producing area of Colombia for nearly 10 years, and the annual output of the emerald ore is declining, and the annual price of the company that makes the emerald raw ore is steadily rising. Raw ore has high quality, but there are also low quality. But if the average price is rising, then the selected boutique emerald will increase in multiple prices in the next few years. Sure enough, in 2023, the price of 5 carats of high-quality emeralds cost 1.2 million yuan. She made money out of this investment, and invited me to eat a very fancy dinner in Beijing.

4. 祖母绿拍卖纪录
Emerald Auction Record

时间 Time	拍卖行 Auction House	拍品 Item	成交价 Closing Price
2024年1月 January, 2024	苏富比香港 Sotheby's Hong Kong	7.13克拉 天然赞比亚无油祖母绿 配 钻石戒指 7.13 carat natural Zambian oil-free emerald with diamond ring	成交价 279,400 港元 Sale Price HKD 279,400
2023年5月 May, 2023	佳得士日内瓦 Christie's Geneva	HARRY WINSTON 17.43克拉哥伦比亚微油祖母绿戒指 HARRY WINSTON 17.43 carat Colombian emerald ring with minor-oil	成交价 1,799,500法郎 Sale Price CHF 1,799,500
2022年7月 July, 2022	苏富比香港 Sotheby's Hong Kong	12.39克拉「哥伦比亚」祖母绿配钻石戒指 12.39 carats "Colombian" emerald with diamond ring	成交价 176,400港元 Sale Price HKD 176,400
2021年12月 December, 2021	佳得士纽约 Christie's New York	6.729克拉哥伦比亚中油祖母绿戒指 6.729 carat Colombian Oil emerald Ring moderate amount of oil	成交价 27,500美元 Sale Price USD 27,500

5. 祖母绿投资收藏建议
Emerald Investment Collection Suggestions

小雨老师 经验分享
Experience Sharing from Ms. Xiaoyu

(1) 关注产地和来源
Focus on the Place of Origin and Sources

最好是选择哥伦比亚产的高品质祖母绿，其次是赞比亚产的。

It is best to choose high-quality emeralds from Colombia, followed by those from Zambia.

(2) 权威证书
Certificate of Authority

购买祖母绿时，确保它配有权威的宝石证书和鉴定报告。证书提供宝石的详细信息，包括重量、颜色、净度和切割等。它们也可以作为宝石真实性和品质的证明。

When buying an emerald, make sure it comes with an authoritative gemstone certificate and identification report. The certificate provides detailed information about the gemstone, including weight, color, clarity and cut. They also serve as proof of the authenticity and quality of the stone.

(3) 保管和维护
Custody and Maintenance

祖母绿避免暴露在强烈的阳光下，避免接触化学物质和硬物。

Emeralds should avoid exposure to strong sunlight and avoid contact with chemicals and hard objects.

(4) 了解市场趋势
Understand Market Trends

可以看一些国外矿业公司的原石产量和拍卖价格，或者咨询专业的名贵珠宝资产配置顾问，他们会有详细的数据分析。

You can look at the raw stone production and auction prices of some foreign mining companies, or consult a professional luxury jewelry asset allocation consultant, they will have detailed data analysis.

祖母绿投资收藏目前首选是哥伦比亚的祖母绿，颜色要达到VVG或者木佐颜色。油量尽量选择微油及极微油，无油就更好，但至少是微油。在净度上尽量选择没有开放性裂隙的祖母绿，最好是用放大镜放大观察一下外部的裂隙情况。在重量上，我们尽量选择3克拉以上的，在自己预算比较宽裕的情况下，越大越好。在证书选择上，我们至少选择GRS证书，比GRS更好的像古柏林、AGL，他们会对祖母绿有一个整体评分，评分越高，它的价值就越高。目前高品质的哥伦比亚祖母绿，4克拉也有可能达到100万元人民币左右，如果是10克拉以上，有可能达到上千万元人民币。

The emerald investment collection is currently preferred to Colombian emeralds, the color to achieve VVG or musa color. Try to choose minor oil and very insignificant oil. No oil is better, but at least minor oil. In terms of clarity, try to choose emeralds without open cracks, it is best to use a magnifying glass to observe the external cracks. In terms of weight, we try to choose more than 3 carats, in the case of your own budget is relatively generous, the larger the better. In the certificate selection, we at least choose GRS certificate, what would be better is Gubelin and AGL. They will have an overall score for the emerald, the higher the score, the higher its value. At present, high-quality Colombian emeralds, 4 carats may also reach about 1 million yuan, if it is more than 10 carats, it may reach tens of millions of yuan.

Alexandrite

第七章
07 Chapter Seven

亚历山大变石
Alexandrite

一、亚历山大变石的基本特征
Basic Characteristics of Alexandrite

亚历山大变石（Alexandrite）是一种非常独特和珍贵的宝石，在拍卖市场上非常受欢迎。白天它吸收日光呈绿色调，晚上它吸收暖光呈黄色或者红色调。它以其特殊的颜色变化而闻名，根据光源的不同，它可以呈现出绿色、蓝色、红色和紫色等多种颜色。亚历山大变石的颜色变化效应令人着迷，使其成为收藏家和珠宝爱好者的宠儿。其独特的光学性质使其在不同光照条件下呈现出不同的颜色，这使得每一颗亚历山大变石都独一无二。

Alexandrite is a very unique and precious gemstone that is very popular on the auction market. During the day it absorbs daylight in a green tone, and at night it absorbs warm light in a yellow or red tone. It is known for its special color variations, which can appear in a variety of colors such as green, blue, red, and purple depending on the light source. The color changing effect of Alexandrite is fascinating, making it a favorite choice among collectors and jewelry lovers. Its unique optical properties make it appear different colors under different lighting conditions, which makes each Alexandrite unique.

The Basic Characteristics of Alexandrite

亚历山大变石

亚历山大变石

小雨老师在泰国考察

小雨老师在矿区

小雨老师在矿区

　　亚历山大变石最早于1834年在俄罗斯的乌拉尔山脉的Emerald Mine被发现，当时正值俄罗斯沙皇亚历山大二世的加冕庆典，因此以沙皇名字命名。亚历山大变石的产量非常有限，并且非常罕见，因此被认为是世界上最昂贵的宝石之一。

　　Alexandrite was first discovered in the Emerald Mine in the Ural Mountains of Russia in 1834 during the coronation celebrations of Tsar Alexander II of Russia, so it was named after the Tsar. Alexandrite is produced in very limited quantities and is so rare that it is considered as one of the most expensive gemstones in the world.

Alexandrite

二、如何鉴赏和投资收藏亚历山大变石
How to Appreciate and Invest in Alexandrite Collection

我们主要从以下 7 个维度去鉴赏和投资亚历山大变石，我在每一个鉴赏指标分析里都给出了目前的市场情况以及相应指标的价格影响，也参考了一些拍卖价格，大家可以参考。

We mainly appreciate and invest Alexandrite from the following seven aspects. In the analysis of each appreciation indicator, I have given the current market situation and the price impact of the corresponding indicators, and also referred to some auction prices for your reference.

（一）亚历山大变石的鉴赏
Appreciation of Alexandrite

1. 颜色
Color

　　亚历山大变石的质量主要取决于变色效应的明显程度以及在不同光下的颜色纯度和艳丽程度。亚历山大变石，在白光和暖光下会呈现不同的颜色，不同产地也会呈现不同的颜色，而且变色越明显越稀有。

　　The quality of Alexandrite depends mainly on the degree of discoloration effect and the degree of color purity and showiness under different light. Alexandrite stone, under white and warm light sources will show different colors, different origins will show different colors, and the more obvious the discoloration, the rarer.

淡绿色—褐红色
Light Green - Maroon

一般来自斯里兰卡，高品质。
Generally from Sri Lanka, high quality.

淡绿色—褐红色亚历山大变石

淡黄绿色—褐黄红色
Light Yellow/Green - Brown Yellow-Red

一般来自斯里兰卡。
Generally from Sri Lanka.

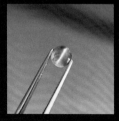 

淡黄绿色—褐黄红色亚历山大猫眼

蓝绿色—紫红色
Turquoise - Fuchsia

一般来自俄罗斯，高品质，很贵。
Generally from Russia, high quality, very expensive.

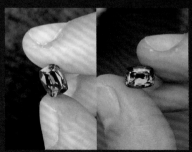

蓝绿色—紫红色亚历山大变石

绿色—红色
Green - Red

一般来自巴西，高品质，非常贵。
Generally from Brazil, high quality, very expensive.

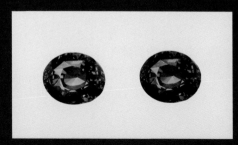

绿色—红色亚历山大变石

小雨老师经验分享
Experience Sharing from Ms. Xiaoyu

亚历山大变石的价格取决于它颜色变化的强烈程度,比如说它从红色到绿色,变化得特别强烈,那么价格就特别高,但如果是从棕色调儿的红色变为黄色调儿的绿色,那么它的价格就会相对低一些,因为颜色纯度越高,价格越高,变化越大,价格越高。打个比方,一颗 5 克拉的亚历山大变石,如果它的绿色跟红色变化得特别明显,它的价格远远高于一颗 10 克拉的颜色变化没有那么明显的亚历山大变石。2012 年,一颗 15 克拉的变色效应特别明显的亚历山大变石拍卖价格是 800 万元人民币左右。

The price of Alexandrite depends on the intensity of its color change, for example, it changes particularly strongly from red to green, then the price is particularly high, but if it changes from brown color red to yellow color green, then its price will be relatively low. This is because the higher the color purity, the higher the price; the greater the change, the higher the price. For example, a 5-carat Alexandrite with a particularly pronounced change in green and red will cost much more than a 10-carat Alexandrite with a less pronounced change in color. In 2012, the auction price of a 15-carat Alexandrian stone with a particularly obvious color-changing effect was about 8 million yuan.

2.净度
Clarity

是指宝石的内外的纯净程度,是否含有包裹体。
This refers to the degree of purity of the stone inside and outside, whether it contains inclusions.

① 包裹体的大小、数量和位置
Size, Number and Location of Inclusions

包裹体越大、数量越多,对宝石的净度影响就越大。特别是当包裹体位于台面部位时,对宝石净度的影响更为显著。

The larger and more inclusions, the greater the impact on the clarity of the stone. Especially when the inclusion is located in the mesa, the influence on the clarity of the gem is more significant.

② 包裹体的类型和对比度
Type and Contrast of Inclusions

包裹体的颜色与宝石的差异也会影响宝石的净度。如果包裹体的颜色与宝石相差较大,对宝石的净度影响会更明显。

The difference between the color of the inclusion and the stone can also affect the clarity of the stone. If the color of the inclusion is different from that of the gem, the impact on the clarity of the gem will be more obvious.

③ 未愈合的裂隙
An Open Fissure

如果宝石内部有未愈合的裂隙,将降低宝石的耐久性,容易受到损伤。

If there are unhealed cracks inside the stone, the durability of the stone will be reduced and it will be vulnerable to damage.

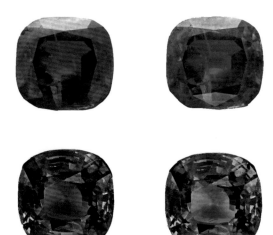

亚历山大变石净度对比

在选择亚历山大变石时，尽量选择表面没有矿坑、明显色带、裂隙和深色包裹体的宝石。只要肉眼或者社交距离没有太影响美观，都是可以接受的。原则上，只要肉眼难以看到宝石内部包裹体，即可认为宝石具有良好的净度。并不需要追求完全干净，因为完全干净的宝石反而可能会被怀疑是合成的。

When selecting Alexandrite, try to select stones with no pits, visible bands, cracks and dark inclusions on the surface. As long as the naked eye or social distance does not affect the appearance too much, we are okay with it. In principle, as long as it is difficult to see the internal inclusions of the stone with the naked eye, the stone can be considered to have good clarity. There is no need to aim for complete cleanliness, as a completely clean gemstone may instead be suspected of being synthetic.

Carat Weight

小雨老师经验分享
Experience Sharing from Ms. Xiaoyu

净度对价格的影响远远没有颜色对价格的影响大，只要肉眼观察不影响整体美观即可。

The impact of clarity on the price is far less than the impact of color on the price, as long as there is no obvious impact on the overall beauty from naked-eye observation.

3. 切工
Cut

是指宝石外表切割的对称度和工整度。
It refers to the symmetry and smoothness of the cut of the gem's appearance.

切工主要表现在对称性和抛光质量等方面。对称性的问题指正侧面轮廓的对称性偏差、台面偏心、底尖偏心、亭部膨胀、刻面畸形和刻面尖点不尖等。抛光质量指的是宝石表面的抛光程度。宝石的切工主要是考虑一个空窗问题，也就是台面过大但厚度过薄，就会形成一个空窗效应，中间那一块颜色很淡或看不到颜色，失去了颜色整体的浓郁度。还有一个就是考虑切割过厚的问题，台面的比例过小，整特别厚，那它的火彩会有一点点发黑，没那么闪亮，切割比例就不舒展。

The cutting is mainly manifested in symmetry and polishing quality. The symmetry problems include the symmetry deviation of the side profile, the eccentricity of the mesa, the eccentricity of the bottom tip, the bulge of the pavilion, the facet deformity and the non-cusp of the facet. Polishing quality refers to the degree of polishing of the gem's surface. The cutting of gems is mainly to consider an empty window problem, that is, the countertop is too large but the thickness is too thin, it will form an empty window effect, and the color of the middle piece is very light or the color can not been seen, and the overall richness of the color is lost. Another point is to consider the problem of cutting too thick, the proportion of the platform is too small, the whole is particularly thick, then its fire color will be a little black, not so shiny, the cutting ratio will not stretch.

切工对比图

小雨老师经验分享
Experience Sharing from Ms. Xiaoyu

切割过厚会涉及价值问题，如果这是一颗 5 克拉的亚历山大，但是由于它的台面小，看着像 3 克拉，你就会觉得不划算，因为 5 克拉的价格可能需要 50 万元人民币，但是 3 克拉的只需要 10 万元人民币。如果切得过厚，5 克拉的货，你要用 3 克拉的价格去还价，这样才划算，我们可以将宝石重新切割。

Cutting too thick will involve the value issue, for example, if this is a 5 carat Alexandrite, but because of its small countertop, it looks like a 3 carat. You will feel that it is not worthwhile, as the price of 5 carat may cost 500,000 yuan, but this product looks like 3 carat, which only costs 100,000 yuan. If the cut is too thick, you have to bargain with 3 carats for 5 carats, then it's a good deal, and we can cut the stone again.

---

4. 重量
Weight

亚历山大变石，晶体大，克拉颜色又很好的比较难得。我们宁愿选择颜色变化效应明显的，而不选择只是大克拉的。颜色才是比较值钱的因素。

Alexandrite, with large crystal and very good carat color are relatively rare. We'd rather choose something with a noticeable color change effect than something that's just a big carat. Color is the more valuable factor.

小雨老师经验分享
Experience Sharing from Ms. Xiaoyu

重量对价格的影响：亚历山大变石也是有克拉溢价的，因为它比较稀有。1 克拉的高品质亚历山大变石也可以上拍卖，拍卖价格大概在 15 万元人民币左右。如果能达到 10 克拉以上，价格在 700 万元人民币左右。

The impact of weight on the price: Alexandrite also has a carat premium, because it is relatively rare. 1 carat high-quality Alexandrite can also be auctioned, and the auction price is about 150,000 yuan. If it can reach more than 10 carats, the price is around 7 million yuan.

5. 产地
Place of Origin

1. 俄罗斯 Russia

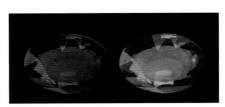
俄罗斯产的亚历山大变石

俄罗斯产的亚历山大变石非常好，它在日光下呈现出类似祖母绿的绿色，在烛光或白炽灯光下呈现出红宝石般的红色。

Russian Alexandrite is very good, it appears emerald-like green in daylight and ruby red in candlelight or incandescent light.

2. 斯里兰卡 Sri Lanka

斯里兰卡产的亚历山大变石

斯里兰卡和南非产的亚历山大变石体色多带有棕色调。

Sri Lankan and South African Alexandrite body color with brown tones.

亚历山大变石在白光和暖光源下会呈现不同的颜色，不同产地也会呈现不同的颜色，而且变色越明显越稀有。

Alexandrite will show different colors under white and warm light sources, and different origins will show different colors, and the more obvious the discoloration, the rarer.

3. 巴西 Brazilian

巴西产的亚历山大变石非常好，变色效应非常明显。在日光灯下呈现出蓝绿或黄绿色，在烛光下呈现紫或紫红色。米纳斯吉拉斯州是巴西优质猫眼和亚历山大变石的主要产地。

Brazilian Alexandrite is very good, the discoloration effect is very obvious. It appears blue-green or yellow-green under fluorescent lamps and purple or purplish red under candlelight. Minas Gerais is Brazil's main producer of high-quality cat's eye and Alexandrite.

巴西产的亚历山大变石

小雨老师经验分享
Experience Sharing from Ms. Xiaoyu

产地对价格的影响：因为亚历山大变石在不同的产地，它的颜色就不同，比如说俄罗斯和巴西的产地，亚历山大变石的变色效应就特别明显，巴西的可以从绿色变到红色，这种强烈的变色效应，价格就非常贵。斯里兰卡产的亚历山大变石，颜色会带一些棕色调和黄色调，所以价格就要便宜很多。1克拉巴西的亚历山大变石，可能价格要10万元人民币以上，但是1克拉斯里兰卡的亚历山大变石，可能只要几千元人民币。

As Alexandrite in different places of origin, its color is different. Like in Russia and Brazil, Alexandrite color change effect is particularly obvious, Brazil can change from green to red. With this strong color change effect, the price is very expensive. Sri Lankan Alexandrite, the color will have some brown and yellow tones, so the price is much cheaper. 1 carat Brazilian Alexandrite may cost more than 100,000 yuan, but 1 gram Sri Lanka Alexandrite may only cost a few thousand yuan.

6. 优化
Optimization

亚历山大变石目前还没有优化。
Alexandrite is not currently optimized.

7. 证书
Certificate

01 GRS

来自瑞士，在证书上会体现亚历山大变石，而且说明产地。

From Switzerland, the certificate will show the Alexandrite and state the origin.

GRS证书

02 SSEF

来自瑞士，在证书上会体现亚历山大变石，而且说明产地。

From Switzerland, the certificate will show Alexandrite and state the origin.

SSEF证书

03 AGL

来自美国，在证书上会体现亚历山大变石，而且说明产地，而且还有颜色的评分。

From the United States, the certificate will show the Alexandrite, and state the origin, and also the score of colors.

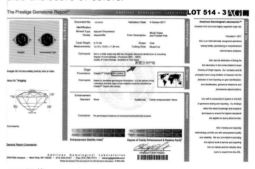

AGL证书

04 Gubelin

来自瑞士，在证书上会体现亚历山大变石，而且说明产地，而且还会有综合的评分。

From Switzerland, the certificate will show the Alexandrite, and state the origin, and there will be a comprehensive score.

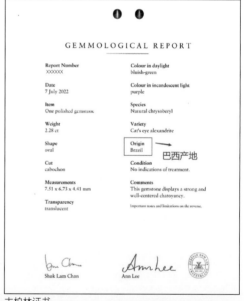

古柏林证书

小雨老师经验分享
Experience Sharing from Ms. Xiaoyu

如果要上拍卖的话，一般需要配有 GRS、SSEF、古柏林、AGL 证书。
If it is to be auctioned, it is generally required to have GRS, SSEF, Gubelin, AGL.

（二）亚历山大变石的投资收藏建议
Alexandrite Investment and Collection Suggestions

1. 亚历山大变石的拍卖纪录
The Auction Record for Alexandrite

时间 Time	拍卖行 Auction House	拍品 Item	成交价 Closing Price
2023年6月 June, 2023	苏富比香港 Sotheby's Hong Kong	5.05克拉 天然巴西未经处理猫眼亚历山大变石 戒指 5.05 carat natural Brazilian untreated cat's eye Alexandrite ring	成交价 444,500港元 Sale Price HKD 444,500
2018年5月 May, 2018	佳士得日内瓦 Christie's Geneva	29.41克拉未经热处理垫形亚历山大变石，锡兰，颜色变化中等 29.41 carat unheated cushioned Alexandrite, Ceylon, medium color variation	成交价 250,000法郎 Sale Price CHF 250,000
2017年11月 November, 2017	佳士得香港 Christie's Hong Kong	3.38ct亚历山大变石，心形切割 3.38ct Alexandrite, heart cut	成交价 600,000港元 Sale Price HKD 600,000
2023年5月 May, 2023	苏富比香港 Sotheby's Hong Kong	亚历山大戒指，椭圆形切割，重2.17克拉，附有SSEF报告编号108062，说明亚历山大铁矿来自巴西，未经热处理。 The Alexandrite ring, oval cut, weighs 2.17 carats and bears SSEF report number 108062, indicating that the Alexandrite iron ore is of Brazilian origin and has not been heat-treated.	成交价 215,900港元 Sale Price HKD 215,900

2.亚历山大变石投资案例
Alexandrite Investment Case

2016 年我参加香港的珠宝展，一个巴西的珠宝商卖给我一颗非常好的亚历山大变石，1 克拉多，价格是 3 万多人民币。2019 年我有个朋友很喜欢，当时我也是缺钱想买房子就把那颗宝石卖了 5 万多人民币，2024 年 3 月我在香港珠宝展看到高品质亚历山大变石在巴西的珠宝商柜台上要卖 15 万元人民币 1 克拉。

In 2016, when I attended a jewelry exhibition in Hong Kong, a Brazilian jeweler sold me a very good Alexandrite gem, 1 carat, at a price of more than 30,000 RMB. In 2019, a friend of mine who liked it very much. At that time, I was also short of money and wanted to buy a house, so I sold that gem for more than 50,000 RMB. In March 2024, I saw high-quality Alexandrite on the counter of Brazilian jewelers at the Hong Kong Jewelry Exhibition, priced at 150,000 RMB per carat.

3.亚历山大变石的投资收藏建议
Alexandrite Investment Collection Advice

(1)权威证书
Certificate of Authority

最好选择经过古柏林、SSEF 或 GRS 等国际权威机构认证的宝石。这些证书可以提供关于宝石的详细信息，包括其品质、重量和特征。

It is best to choose gems certified by an international authority such as Gubelin, SSEF or GRS. These certificates can provide detailed information about the gemstone, including its quality, weight and characteristics.

(2)高品质亚历山大变石
High-quality Alexandrite

优先选择变色效应比较明显的大克拉亚历山大变石。

The large carat Alexandrite with obvious discoloration effect is preferred.

(3)市场趋势
Market Trend

通过国际拍卖或者国际珠宝展了解亚历山大变石市场的供需情况和价格趋势，以便在合适的时机购买或出售。

Learn about supply, demand and price trends in the Alexandrite market through international auctions or international jewelry shows so you can buy or sell at the right time.

小雨老师经验分享
Experience Sharing from Ms. Xiaoyu

亚历山大变石是一个非常小众的投资品，也是宝石收藏者非常喜欢的一款宝石。但是它在大众的流通市场里面很少出现，只有在宝石收藏者之间或者拍卖场上特别受欢迎。如果是想要快速流通的话，我个人不建议去做亚历山大变石的投资，因为大众对它的了解太少了，它没法快速变现，如果快速变现就会亏点钱。但如果是想要长期持有，不着急变现的话，真的可以买大克拉的亚历山大变石，你就等着用时间去换价值，有好的机会就拍卖，会有不错的收益。目前10克拉以上的高品质亚历山大变石，价格在700万元人民币左右，5克拉的亚历山大变石价格在100万元人民币以上。

Alexandrite is a very niche investment and a favorite choice among gem collectors. However, it rarely appears in the general circulation market, and is especially popular among gem collectors or auction houses. If you want to circulate quickly, I personally do not recommend to do Alexandrite investment, because the public knows too little about it, it can not be quickly realized, and if you do realization very quick, you will lose money. But if you want to hold it for a long time and do not hurry to cash in, you can really buy large carat Alexandrite. You will be just waiting to use time to change value. When there is a good opportunity to auction, there will be a good yield. At present, the price of high-quality Alexandrite of more than 10 carats is about 7 million yuan, and the price of 5-carat Alexandrite is more than 1 million yuan.

亚历山大变石里面有一个特殊的品种叫亚历山大变石猫眼，它是猫眼，但它也是亚历山大变石，只是在变石的基础上，它又多了一个猫眼的现象。像这样具有特殊光学效应的品种，你会觉得我买了一颗宝石相当于买了两颗宝石，尤其是变色效果非常明显、猫眼效果也特别明显的，它在拍卖场上非常受欢迎，真的算是一颗难求。这种极品的货，已经不是买方市场去评估价格了，而是卖方市场掌握了议价权。

Alexandrite has a special breed called Alexandrite cat's eye. It is cat's eye, but it is also Alexandrite. It is on the basis of Alexandrite, and it has a cat's eye phenomenon. With a special optical effect like this, you will feel that I bought a gem equivalent to buying two gems, especially the color change effect is very obvious, and the cat's eye effect is particularly obvious. It is very popular in the auction, and is really hard to find. This kind of excellent goods, is no longer the buyer's market to evaluate the price, but the seller's market who has the bargaining power.

Golden Green Cat's Eye

第八章
08 Chapter Eight

金绿猫眼
Golder Green
Cat's Eye

一、金绿猫眼的基本特征
Basic Characteristics of Golden Green Cat's Eye

金绿猫眼简称猫眼，英文名是Cat's Eye。它是一种具有独特光学特性的宝石，其猫眼效应使其在珠宝市场中备受追捧。它呈现出类似猫眼的狭缝效果，这是由于宝石内部的微小纤维或纤维状包体的存在，使光线在宝石上产生反射和折射。金绿猫眼在拍卖场上比较小众，但是很受日本藏家的欢迎。其实猫眼有很多，碧玺也有猫眼，祖母绿也有猫眼，但是只有金绿猫眼，我们在命名的时候叫做猫眼，其他的都叫做什么什么猫眼，比如海蓝宝猫眼。猫眼在古罗马时就已有记载，斯里兰卡和印度曾是主要的产地。然而，在数个世纪后，金绿猫眼似乎被人们遗忘，直到1879年，英国维多利亚女王的第三个儿子康诺特公爵将一颗金绿猫眼宝石戒指作为订婚戒指赠送给普鲁士公主路易丝·玛格丽特，重新引起了人们的关注。金绿猫眼的独特性和历史背景赋予了它特殊的价值和吸引力，使其成为珠宝收藏家和爱好者追逐的对象。

The golden green Cat's Eye, simply called cat's eye. It is a gemstone with unique optical properties, and its cat's eye effect makes it highly sought after in the jewelry market. It presents a slit effect similar to a cat's eye, due to the presence of tiny fibers, or fibrous inclusions, inside the gem that reflect and refract light on the gem. Cat's eyes in gold and green are relatively rare at auction, but are popular with Japanese collectors. In fact, there are a lot of cat eyes, tourmaline also has cat eyes, emeralds also have cat eyes. But only golden green cat eyes, we are called cat eyes when naming, other are called '**' cat eyes, such as aquamarine cat eyes. Cat's eyes have been recorded in ancient Rome, and Sri Lanka and India were the main sources of production. For centuries, however, the golden and green cat's eye seemed to be forgotten until 1879, when the Duke of Connaught, the third son of Queen Victoria, gave a golden and green cat's eye ring as an engagement ring to Princess Louise Margaret of Prussia, bringing it back to the spotlight. The uniqueness and historical background of the golden green cat's eye give it special value and appeal, making it the object of pursuit by jewelry collectors and enthusiasts.

Basic Characteristics of Golden Green Cat's Eye

金绿猫眼耳钉

小雨老师矿区考察

小雨老师在矿区

金绿猫眼戒指

小雨老师在矿区

Basic Characteristics of Golden Green Cat's Eye

Golden Green Cat's Eye colors primarily fall into two color categories series: yellow and green.

Cat's Eye

二、如何鉴赏和投资收藏金绿猫眼
How to Appreciate and Invest in the Golden Green Cat's Eye Collection

我们主要从以下7个维度去鉴赏和投资一颗宝石，我在每一个鉴赏指标分析里都给出了目前的市场情况以及相应指标的价格影响，也参考了一些拍卖价格，大家可以参考。

We mainly appreciate and invest a gem from the following seven aspects. In each appreciation index analysis, I have given the current market situation and the price impact of the corresponding indicators, and also referred to some auction prices for your reference.

（一）金绿猫眼的鉴赏
Appreciation of Golden Green Cat's Eye

1. 颜色
Color

是指宝石有哪些颜色。
It refers to the color of the stone.

金绿猫眼
金绿猫眼的颜色主要分两个色系，一个是黄色系，一个是绿色系。

The color of the golden green cat's eye is mainly two-color systems, one is yellow, and the other is green.

黄色系
Yellow Series

黄色系中会有浅黄色、棕黄色、黄棕色、黄蜜糖色，等等。比较受欢迎的颜色是蜜糖色，是一种金黄金黄的颜色，尤其受到日本人的喜欢，很像猫的眼睛。

Yellow series In the yellow series, there will be light yellow, brownish yellow, yellow brown, yellow honey, and so on. One of the more popular colors is honey, a golden yellow color that is especially favored by the Japanese and resembles the eyes of a cat.

绿色系
Green Series

绿色系的金绿猫眼有浅绿、中绿、黄绿。但是一般说到绿色系的时候，我们第一反应想到的是亚历山大变石猫眼。它吸收白光的时候是绿色的，吸收暖光的时候会变成红色调。这种亚历山大变石猫眼，虽然很小众，但是值得投资收藏。目前在市面上拍卖的亚历山大变石猫眼，价格可以达到每克拉几十万元人民币。

Green Series The golden green cat's eyes of the green series are light green, medium green, yellow green. But when it comes to green, the first thing we think of is Alexandrite cat's eye. It is green when it absorbs white light and turns red when it absorbs warm light. This Alexandrite cat eye, although very small, but worth investing in the collection. At present, the Alexandrite cat's eye is auctioned on the market, and the price can reach hundreds of thousands of yuan per carat.

小雨老师经验分享
Experience Sharing from Ms. Xiaoyu

金绿猫眼颜色对价格的影响非常大，颜色变化效应越大，价格越高，颜色变化效应越小，价格越低。目前最高的颜色变化是从绿色到红色，这样的变化目前价格是最高的，也是最受拍卖市场欢迎的，能每克拉达到几十万元。

The effect of cat's eye color on the price is very large, the greater the color change effect, the higher the price, the smaller the color change effect, the lower the price. At present, the highest color change is from green to red, and this change is currently the highest price and the most popular in the auction market, which can reach hundreds of thousands of yuan per carat.

黄色系猫眼

绿色系猫眼

Cat's Eye

2.净度
Clarity

是指宝石的内外的纯净程度，是否含有包裹体。
It refers to the degree of purity of the stone inside and outside, whether it contains inclusions.

1 包裹体的大小、数量和位置
Size, Quantity and Location of Inclusions

包裹体越大、数量越多，对宝石的净度影响就越大。特别是当包裹体位于台面部位时，对宝石净度的影响更为显著。
The larger and more inclusions, the greater the impact on the clarity of the stone. Especially when the inclusion is located in the mesa, the influence on the clarity of the gem is more significant.

2 包裹体的类型和对比度
Type and Contrast of Inclusions

包裹体的颜色与宝石本身的颜色很相近的话，对宝石的净度影响不大。如果包裹体的折射率和颜色与红宝石相差较大，对宝石的净度影响会更明显。
If the color of the inclusion is very close to the color of the stone itself, it has little effect on the clarity of the stone. If the refractive index and color of the inclusion are significantly different from the ruby, the impact on the clarity of the gem will be more obvious.

3 未愈合的裂缝
An Open Fssure

如果宝石内部有未愈合的裂隙，将降低宝石的耐久性，容易受到损伤。
If there are unhealed cracks inside the stone, the durability of the stone will be reduced and it will be vulnerable to damage.

其实在实际生活中，肉眼或者社交距离看不到的明显瑕疵，被认为是可接受的范围。然而，带有明显瑕疵并且影响宝石整体美观的宝石会被认为是劣质的。当然，完全无瑕疵的金猫绿眼是最理想的，但这种品质的宝石非常罕见。大部分的绿眼都会有一些内含物或外部瑕疵存在，只要这些瑕疵不明显且不影响整体美感，仍然可以被认为是高品质的宝石。

In fact, in real life, obvious flaws that are invisible to the naked eye or social distance are considered acceptable. However, stones with obvious flaws that affect the overall beauty of the stone are considered inferior. Of course, a completely flawless golden cat-green eye is ideal, but gemstones of this quality are very rare. Most green eyes will have some inclusions or external flaws present, and as long as these flaws are not obvious and do not affect the overall aesthetic, they can still be considered high-quality stones.

小雨老师经验分享
Experience Sharing from Ms. Xiaoyu

净度较差的猫眼

净度较好的猫眼

金绿猫眼之所以能形成猫眼效应，是因为它内部含有定向排列的针状包裹体，形成又亮又直的眼线。我们在评价金绿猫眼的时候，通过眼线来判断它的价格，眼线越亮越直，价格越高，眼线弯弯曲曲的，没那么亮，价格就会低。金绿猫眼的净度主要是考虑它有没有裂隙，尤其是那种开放性的裂隙。如果这个裂隙在底部，不影响整体的美观，我们是可以接受的，但如果有些裂隙就在猫眼效应的台面儿上，就特别影响美观，价格会大打折扣。

The reason why the golden green cat eye can form the cat eye effect is that it contains needle-like inclusions in a directional arrangement, forming a bright and straight eyeliner. When we evaluate the golden green cat's eye, we judge its price through the eyeliner, the brighter and straighter the eyeliner, the higher the price; with the curved eyeliner, not so bright, the price will be low. The clarity of the cat's eyes is mainly to consider whether it has cracks, especially the kind of open cracks. If the crack is at the bottom and does not affect the overall beauty, we can accept it, but if some cracks are on the countertop of the cat's eye effect, it will especially affect the beauty, and the price will be greatly discounted.

3.切工
Cut

指宝石外表切割的对称度和工整度。
It refers to the symmetry and fineness of the cut of the gem.

切工主要表现在，对称性和抛光质量等方面。

The cutting is mainly manifested in the aspects of symmetry and polishing quality.

切工较差的猫眼

切工较好的猫眼

小雨老师经验分享
Experience Sharing from Ms. Xiaoyu

金绿猫眼的切工主要考虑它的弧面的弧度是否完美，越饱满的金绿猫眼价格越高，如果是扁扁的，金绿猫眼的弧度不够完美的话，价格就比较低。就像猫的眼睛一样，整个鼓鼓的，亮亮的，那么它的价格就要高很多，同样是5克拉，我们宁愿买鼓鼓的，亮亮的，也不要买那种扁扁的。

The cutting of the golden green cat eye mainly considers whether the curvature of its curved surface is perfect, the fuller the golden green cat eye the higher the price, if it is flat, the arc of the golden green cat eye is not perfect, the price is relatively low. Just like the cat's eyes, the whole puffy, shiny, then it's a lot more expensive. As for 5 carats, we'd rather buy the puffy, shiny, rather than buy the flat one.

4.克拉重量
Carat Weight

一般来说，金绿猫眼可以长得比较大，有10克拉、20克拉、30克拉的，我们想要收藏一颗金绿猫眼的话，尽量选择克拉数大

Generally speaking, the golden green cat eye can grow relatively large, there are 10 carats, 20 carats, 30 carats. If we want to collect a golden green cat eye, try to choose the larger carats.

猫眼项链

小雨老师经验分享
Experience Sharing from Ms. Xiaoyu

从整个名贵宝石的结构来看，金绿猫眼是目前在市场上严重被低估的一个品类，而且克拉溢价不像其他宝石克拉溢价差距那么大。打个比方，红宝石可能5克拉的现在要300万元人民币，但是10克拉的红宝石要2500万元人民币。但是金绿猫眼10克拉和20克拉的价差是没有那么大的。作为一个小众的投资收藏品，大家可以多关注一下金绿猫眼这个品类。

From the perspective of the structure of the entire luxury stone, the golden green cat's eye is currently a category that is seriously undervalued in the market, and the carat premium is not as large as the carat premium gap of other gems. For example, a 5 carat ruby may cost 3 million yuan, but 25 million yuan for a 10 carat ruby. But the price difference between 10 and 20 carats is not that big. As a niche investment collection, you can pay more attention to the golden green cat eye category.

5. 产地
Place of Origin

是指金绿猫眼的出产地，对价格影响挺大的。
It refers to the origin of the golden green cat's eye, which has a great impact on the price.

① 斯里兰卡 / Sri Lanka

斯里兰卡是金绿猫眼的重要产地之一。这里的金绿猫眼宝石以其高品质和较大的尺寸而闻名。

Sri Lanka is one of the important sources of golden green cat's eye. The golden green cat's eye stones here are known for their high quality and large size.

斯里兰卡猫眼

② 巴西 / Brazil

巴西也是金绿猫眼的一个重要产地。这里的金绿猫眼宝石在颜色和品质上有一定的变化，包括浅绿色到深绿色的范围。巴西产的亚历山大变石猫眼是非常稀有的品种。

Brazil is also an important source of golden green cat's eyes. The golden green cat's eye gemstones here have a certain variety in color and quality, including the range of light green to dark green. The Alexandrite cat eye from Brazil is a very rare breed.

巴西猫眼

③ 缅甸 / Myanmar

缅甸是世界上许多宝石的主要产地之一，包括金绿猫眼。缅甸的金绿猫眼宝石以其独特的颜色和光学效应而著名。

Myanmar is one of the world's leading producers of many luxury stones, including the golden green cat's eye. The golden green cat's eye gemstones of Myanmar are famous for their unique color and optical effects.

缅甸猫眼

小雨老师经验分享
Experience Sharing from Ms. Xiaoyu

在五大国际宝石中，金绿猫眼的产地对它的价格影响是最小的。打个比方，一颗10克拉的哥伦比亚祖母绿大概价格是大几百万元人民币，而一颗赞比亚的10克拉祖母绿价格可能只需要几十万元人民币，所以说产地对这个品种的价格影响特别大。但是金绿猫眼，不论是斯里兰卡的还是巴西的，价差影响没那么大。但如果是巴西产的亚历山大变石猫眼，要比普通的金绿猫眼价格高很多，因为它集合了变色跟猫眼两种特殊的光学效应。目前10克拉的金绿猫眼价格大概在中几十万元人民币，但如果是10克拉的亚历山大变石猫眼，价格都是过百万元人民币的。

Among the five major international gemstones, the origin of the golden green cat's eye has the least impact on its price. For example, the approximate price of a 10-carat Colombian emerald is several million yuan, while the price of a 10-carat Zambian emerald may only need hundreds of thousands of yuan, so the origin has a particularly large impact on the price of this variety. But the gold and green cat's eye, whether Sri Lanka's or Brazil's, has less of an impact. However, if it is a Alexandrite cat's eye, the price is much higher than the ordinary golden green cat eye, because it combines two special optical effects of discoloration and cat eyes. At present, the price of 10 carat gold and green cat eyes is probably hundreds of thousands of yuan, but if it is 10 carat Alexander Alexandrite cat eyes, the price is more than one million yuan.

6. 优化
Optimize

目前金绿猫眼还没有优化工艺出现。
At present, the golden green cat eye has not been optimized.

7. 证书
Certificate

不同的证书代表不同的含金量。
Different certificates represent different gold .

01
GRS

证书上会标明金绿猫眼的产地、颜色等基本特征。

GRS certificate: The certificate will indicate the origin of the golden green cat's eye, color and other basic characteristics.

02
Gubelin

证书上会标明金绿猫眼的产地、颜色等基本特征，会给金绿猫眼打一个整体评分。

Gubelin certificate: The certificate will indicate the origin of the golden green cat eye, color and other basic characteristics, and will give an overall score for the golden green cat eye.

GRS证书

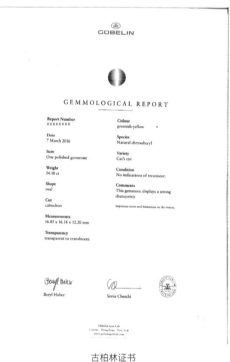

古柏林证书

小雨老师经验分享
Experience Sharing from Ms. Xiaoyu

我们在收藏一颗金绿猫眼的时候，最基本的是要拿到GRS的证书，颜色应该是蜜糖色为最佳。如果是想要一个收藏级的金绿猫眼，我个人建议古柏林证书会更好，因为它不但会对产地、颜色有评级，还会针对这颗金绿猫眼的整体打一个评分，评分越高，证明这颗金绿猫眼的质量就越好。

When we collect a golden green cat's eye, the most basic thing is to get the GRS certificate, and the color should be honey color as the best. If you want a collection level gold and green cat's eye, I personally suggest that the Gubelin certificate will be better, because it will not only have a rating for the origin, color, but also for the overall score of the gold and green cat's eye, the higher the score, the better the quality of the gold and green cat's eye.

(二) 金绿猫眼的投资收藏建议
Golden Green Cat's Eye Investment and Collection Suggestions

1. 金绿猫眼拍卖纪录
Golden Green Cat's Eye Auction Record

时间 Time	拍卖行 Auction House	拍品 Item	成交价 Closing Price
2023年11月 November, 2023	佳士得香港 Christie's Hong Kong	20.31 克拉天然 斯里兰卡金绿猫眼戒指 20.31 carat natural Sri Lankan cat's eye ring in Golden green	成交价 163,800 港元 Sale Price HKD 163,800
2023年4月 April, 2023	苏富比香港 Sotheby's Hong Kong	9.91克拉天然 斯里兰卡金绿猫眼戒指 9.91 carat natural Sri Lankan cat's eye ring in golden green	成交价 95,250 港元 Sale Price HKD 95,250
2021年10月 October, 2021	佳士得香港 Christie's Hong Kong	7.25ct金绿猫眼戒指 7.25ct golden green cat is eye ring	成交价 27,500港元 Sale Price HKD 27,500
2020年7月 July, 2020	苏富比香港 Sotheby's Hong Kong	12.03克拉天然 斯里兰卡金绿猫眼石 配 白水晶 及 钻石 戒指 12.03 carat natural Sri Lankan golden green cat's eye with white crystal and diamond ring	成交价 475,000 港元 Sale Price HKD 475,000

2.猫眼的投资案例
Golden Green Cat's Eye Investment Case

2015年我还在上学时,跟老师一起去斯里兰卡宝石矿区考察,当时就被一颗10克拉的金绿猫眼吸引了,那个时候的价格才3万元人民币,现在10克拉以上的高品质金绿猫眼至少是10万元人民币以上了,金绿猫眼是所有名贵珠宝中涨价最少的。

In 2015, when I was still in school, I went to Sri Lanka with my teacher to visit the gem mining area and I was attracted by a 10-carat gold and green cat eye. At that time, the price was only 30,000 yuan, and now the high-quality gold and green cat eye over 10 carats is at least 100,000 yuan, which is the least price increase among all luxury jewelry.

3.猫眼的投资收藏建议
Golden Green Cat's Eye Investment and Collection Suggestions

(1) 权威证书:选择名贵珠宝投资顾问或者专业度高可信度高的珠宝商。最好选择经过古柏林、SSEF或GRS等国际权威机构认证的宝石。这些证书可以提供关于宝石的详细信息,包括其品质、重量和特征。

Authoritative Certificate: Choose a luxury jewelry investment adviser or a professional jeweler with high credibility. It is best to choose gems certified by an international authority such as Gubelin, SSEF or GRS. These certificates can provide detailed information about the gemstone, including its quality, weight and characteristics.

(2) 高品质猫眼:优先选择10克拉以上的蜜糖色、净度好、眼线亮又直的金绿猫眼。

High-Quality Cat's Eye Gemstones: Preference for 10 carats or more honey color, good clarity, bright and straight eyeliner of golden green cat's eyes.

(3) 市场趋势:多了解国际名贵珠宝价格趋势,如果您对金绿猫眼的投资收藏不太了解或感到困惑,建议寻求专业的珠宝投资顾问的建议。

Market Trends: Learn more about international luxury jewelry price trends, and if you are not familiar with or confused about the gold and green cat's eye investment collection, it is recommended to seek the advice of a professional jewelry investment adviser.

小雨老师 经验分享
Experience Sharing from Ms. Xiaoyu

金绿猫眼是一个特别小众的投资品,它比较受日本人和欧美人的欢迎,在国内市场上大多数收藏者对金绿猫眼的认知还比较浅。如果想要收藏的话,最好选择10克拉以上的金绿猫眼,大概价格在几十万元人民币,作为投资的话,我个人建议选亚历山大变石猫眼,这种非常稀有的品种,未来会更具有投资价值。

Golden green cat's eye is a particularly niche investment goods, it is more popular with the Japanese, Europe and the United States. In the domestic market, most collectors' recognition on the golden green cat's eye is relatively low. If you want to collect it, it is best to choose more than 10 carats of golden green cat's eyes, the price is probably hundreds of thousands of yuan. As an investment, I personally recommend choosing Alexandrite cat's eyes, this very rare variety, and in the future there will have more investment value.

Black Opal

第九章
09 Chapter Nine

黑欧泊
Black Opal

黑欧泊作为一种珍贵而稀有的宝石，具有独特的美丽和变彩效应，被珠宝大品牌广泛应用于高端珠宝设计中。卡地亚是世界知名的珠宝品牌，他们在一些限量版和高级珠宝系列中使用了黑欧泊，例如，他们的高级珠宝项链和戒指中常常会使用黑欧泊作为中心宝石，或作为镶嵌宝石的一部分。蒂凡尼也是世界顶级珠宝品牌之一，他们在一些珠宝系列中使用了黑欧泊，例如，他们的蓝盒系列中的一些戒指和耳环可能会镶嵌有黑欧泊，增添了珠宝的华丽和独特性。梵克雅宝以其细腻而精致的珠宝设计闻名，他们也应用了黑欧泊在一些高级珠宝系列中，例如，他们的Alhambra系列中的一些项链、手链和耳环会使用黑欧泊作为主要宝石，呈现出迷人的光彩。宝格丽在他们的高级珠宝系列中也使用了黑欧泊，例如，他们的Divas' Dream系列中的一些项链、戒指和耳环可能会镶嵌有黑欧泊，以展现出该品牌独特的意大利风格和奢华感。

As a precious and rare gemstone, Black Opal has a unique beauty and color changing effect, which is widely used in high-end jewelry design by major jewelry brands. Cartier is a world-renowned jewelry brand, and they use black opal in some limited edition and fine jewelry collections, for example, their fine jewelry necklaces and rings often use black opal as a center stone, or as part of the setting stone. Tiffany is also one of the top jewelry brands in the world, and they use black opal in some of their jewelry collections, for example, some rings and earrings in their blue box collection may be inset with black opal, adding to the gorgeousness and uniqueness of the jewelry. Van Cleef & Arpels is known for its delicate and sophisticated jewelry designs, and they also apply black opal in some of their fine jewelry collections, for example, some necklaces, bracelets and earrings in their Alhambra collection, using black opal as the main stone, presenting a charming glow. Bulgari also uses black opal in their fine jewelry collection, for example, some necklaces, rings and earrings in their Divas' Dream collection, which inset with black opal to show off the brand's unique Italian style and sense of luxury.

一、黑欧泊的基本特征
Basic Characteristics of Black Opal

黑欧泊英文名：Black Opal。在深色的胚体色调上呈现出明亮色的有彩蛋白石，我们称之为黑欧泊。它是欧泊的一种，欧泊就是蛋白石，有特别多的品种，铁欧泊、黑欧泊、水晶欧泊、火欧泊，等等，黑欧泊是所有的欧泊品种当中，价值最高的。黑欧泊出产于澳大利亚新南威尔士州闪电岭，天然黑欧泊是欧泊中的皇族，由于他们的形态和稀有，高品质黑欧泊价格昂贵。

Black Opal, the opal with bright color out of a dark embryo tone. There are many varieties of opal, iron opal, black opal, crystal opal, fire opal, etc., Black pal is the highest value one among all the Opal varieties. Black Opal is produced in Lightning Ridge, New South Wales, Australia, natural black opal is the royal family of opal. Due to their form and rarity, high quality black opal is expensive.

小雨老师宝石交易

欧泊胸针

小雨老师宝石交易

欧泊戒指

  黑欧泊在拍卖场上也是非常受欢迎的，在拍卖场上的价格不断上涨。2019 年，一枚重达 14.62 克拉的黑欧泊在澳大利亚卡塔尔宝石拍卖会上以 300 万澳元成交，这颗宝石被誉为"星之子"，因其深邃的黑色背景和丰富的彩虹色变彩而备受瞩目。2016 年，一颗重达 12.03 克拉的黑欧泊在苏富比拍卖会上以 175 万美元成交，这颗宝石来自澳大利亚闪电岭，有着较大的尺寸和绚丽的变彩效应。

  Black opal is also very popular in the auction house, where prices continue to rise. In 2019, a 14.62-carat black opal, known as the "Child of the Stars" and notable for its deep black background and rich rainbow color variations, was sold for $3 million at the Qatar Gemstone Auction in Australia. In 2016, a 12.03-carat black opal from Lightning Ridge, Australia, with its large size and gorgeous transcolor effect, sold for $1.75 million at Sotheby's.

Black opal

二、如何鉴赏和投资收藏黑欧泊
How to Appreciate and Invest in the Black Opal Collection

我们主要从以下6个维度去鉴赏和投资黑欧泊，我在每一个鉴赏指标分析里都给出了目前的市场情况以及相应指标的价格影响，也参考了一些拍卖价格，大家可以参考。

We mainly appreciate and invest black opal from the following six aspects. In each appreciation index analysis, I have given the current market situation and the price impact of the corresponding indicators, and also referred to some auction prices for your reference.

（一）黑欧泊的鉴赏
Appreciatior of Black Opal

1.颜色
Color

黑欧泊的体色为黑色或深蓝、深灰、深绿、褐色等。但是黑欧泊的表面会呈现五彩斑斓的亮斑，很像一组彩虹。有的欧泊以蓝—绿色为主，有的欧泊以橙色—红色为主，红色变彩越多价值越高。

Black opal body color is black or dark blue, dark gray, dark green, brown, etc. But the surface of the black opal will appear colorful bright spots, much like a group of rainbows. Some opal is mainly blue-green, and some opal is mainly orange-red, and the more red color changes, the higher the value.

橙红色调欧泊　　黄绿色调欧泊

 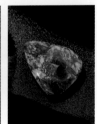

蓝绿色调欧泊　　蓝紫色调欧泊

2. 特征
Features

变彩效应是黑欧泊最大的特征，通常欧泊具有变彩效应，在光源下转动欧泊可以看到五颜六色的色斑。变彩颜色种类越多、越鲜艳，其价值越高，最多能有7种变彩颜色。其中红色变彩越多价值越高

The color change effect is the largest feature of black opal, usually opal has a color change effect. Rotating opal under the light source, you can see colorful spots. The more variety and vividness of the color change, the higher its value, and there can be up to 7 different color change. The more red color changes, the higher the value.

闪电岭欧泊

3. 产地
Origin

黑欧泊的产区在澳大利亚新南威尔士州闪电岭。

Black opal is grown in Lightning Ridge, New South Wales, Australia.

4. 亮度
Brightness

有些黑欧泊表面特别亮，就像一面镜子，但是有些黑欧泊表面就雾蒙蒙的，我们在选择的时候尽量选择亮度高的黑欧泊。从价值上来讲，亮度越高，价值越大。

Some black opal surface is particularly bright, like a mirror, but some black opal surface is foggy. We try to choose high brightness black opal when choosing. In terms of value, the higher the brightness, the greater the value.

亮度优秀欧泊

5. 净度
Clarity

欧泊虽然是不透明的，它不存在内部包裹体，但是表面有裂的欧泊价值要低很多。

Although opal is opaque, it does not have internal inclusions, but the surface of the cracked opal's value is much lower.

七彩欧泊戒指

6. 双面欧泊
Double Opal

大多数欧泊都是单面的，它底层是玛瑙或者脉石，表层就是变彩，但是有的欧泊是可以做到双面变彩的，两面都有变彩效果，这样的欧泊要贵很多。一颗10克拉的双面变彩欧泊，价值过百万元人民币。

Most opal is single-sided, its bottom is agate or gangue, the surface is changing color. But some opal can do double-sided change color, both sides have changing color effect, such opal is much more expensive. A 10-carat double-sided variable color opal, worth more than one million yuan.

双面欧泊吊坠

(二)黑欧泊的投资收藏建议
Black Opal Investment and Collection Suggestions

1.黑欧泊涨价的原因
The Reason for the Price Increase of Black Opal

95%的黑欧泊来自澳大利亚的闪电岭矿区,并且这些原石中95%也只是普通蛋白石,只有不到5%才是有色彩的欧泊,而且大部分都非常小,通常原石都不到5克拉。经过100多年的不断开采,矿产资源几乎枯竭,产量稀少,开采难度也越来越大,开采成本高昂,使得欧泊价格连年上涨。从近20年的珠宝市场来看,欧泊属于价格快速增长的类型。平均下来,欧泊升值的幅度大概在10倍上下,2001年前后售价30万元人民币的50克拉欧泊原料石,在2010年的卖价大约在150万元人民币,折合每克拉3万元人民币。2020年最优质的黑欧泊,售价已高达每克拉35万元人民币以上。

95% of black opal comes from the Lightning Ridge Mine in Australia, and 95% of these raw stones are ordinary opals, less than 5% are colored opals, and most of them are very small, usually less than 5 carats of raw stone. After more than 100 years of continuous mining, mineral resources are almost exhausted. The production is scarce, the mining difficulty is becoming more and more difficult, and the mining cost is high, making the price of opal rise year after year. From the perspective of the jewelry market in the past 20 years, opal belongs to the type of rapid price growth. On average, the appreciation rate of opal is about 10 times, and the 50 carat opal raw material stone with a price of 300,000 yuan around 2001, which was sold at a price of about 1.5 million yuan in 2010, equivalent to 30,000 yuan per carat. The best quality black opal in 2020 has sold for more than 350,000 yuan per carat.

2.黑欧泊拍卖价格
Black Opal Auction Price

时间 Time	拍卖行 Auction House	拍品 Item	成交价 Closing Price
2024年12月 December, 2024	苏富比纽约 Sotheby's New York	卡地亚黑欧泊项链,欧泊大小约为29.2mm×20.9mm×5.6 mm Cartier black opal necklace, opal size is about 29.2mm*20.9mm×5.6mm	成交价 241,300 美元 Sale Price USD 241,300
2023年6月 June, 2023	苏富比纽约 Sotheby's New York	卡地亚黑欧戒指,欧泊大小约为30.0mm×19.7mm×7.8 mm Cartier black European ring, Opal size is about 30.0mm×19.7mm×7.8mm	成交价 13,970 美元 Sale Price USD 13,970
2023年5月 May 2023	邦翰思香港 Bonhams Hong Kong	9.08克拉澳大利亚闪电岭天然未经加热处理黑欧泊配钻石戒指 9.08 carat Australian Lightning Ridge Natural unheated Black Opal with diamond ring	成交价 396,800 港元 Sale Price HKD 396,800

时间 Time	拍卖行 Auction House	拍品 Auction Item	成交价 Current Rate
2022年12月 December, 2022	苏富比纽约 Sotheby's New York	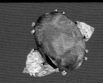 黑欧泊配钻石戒指，欧泊大小约为18.9mm×13.7mm Black Opal with diamond ring, Opal size is about 18.9mm×13.7mm	成交价 68,750 美元 Sale Price USD 68,750

3.黑欧泊投资案例
Black Opal Investment Case

　　我在2016年去日本首饰学院访学的时候，正好遇到东京国际珠宝展，日本是20世纪90年代销售黑欧泊最火热的国家，近几年日本的珠宝展主要卖的就是90年代的黑欧泊。2016年我买了一个很好的3克拉的黑欧泊吊坠，价格是3万元人民币，2024年3月我在香港国际珠宝展上又遇到那个日本的参展商，一颗高品质的3克拉的黑欧泊戒指价格是20万元人民币。

　　When I went to Japan Jewelry Institute to study as a visiting student in 2016, I happened to encounter the Tokyo International Jewelry Exhibition. Japan was the hottest country in the sales of black opal in the 1990s. In recent years, the jewelry exhibition in Japan mainly sold black Opal in the 1990s. In 2016, I bought a very good 3-carat black Opal pendant, the price is 30,000 yuan. In March 2024, I met the Japanese exhibitor at the Hong Kong International Jewelry Fair, and the price of a high-quality 3-carat black Opal ring is 200,000 yuan.

4.黑欧泊的投资收藏建议
Black Opal's Investment And Collection Advice

 小雨老师经验分享
Experience Sharing from Ms. Xiaoyu

　　黑欧泊是一个非常小众的宝石投资品，它在宝石收藏家或者投资者人群里非常受欢迎，但是市面上对它的认知还是非常少，在欧美国家和日本对黑欧泊的认知会多一些，欧洲的古董首饰和一些日本回流的珠宝经常出现黑欧泊，但是品相有好有坏。这两年才在国内慢慢开始对它有认知。我们在选择黑欧泊的时候，第一尽量选择红色系、蓝绿色系，红色系的价值要比蓝绿色系的价值高一些。第二尽量选择亮度好的，因为这样的欧泊变彩效果更好。第三，尽量选择弧度是拱起来的黑欧泊，因为可以让火彩呈现得更好。第四，在自己预算范围内，选择尺寸大一点的黑欧泊。

　　Black opal is a very minority gem investment product, it is very popular among gem collectors or investors, but the market is still having very little awareness of it. In Europe and the United States and Japan, the awareness of black opal is more, European antique jewelry and some Japanese returned jewelries are often using black opal, but the quality is good/bad occasionally. In the past two years, people slowly began to understand it in China. When we choose black opal, the first is to try to choose red series, or blue green series, the value of the red series is higher than the value of the blue green series. Second, try to choose good brightness, because such opal color change effect is better. Third, try to choose the black opal that is arched, because it can make the fire color present better. Fourth, choose a larger black opal within your budget.

The Variety of Color Diamonds

第十章
10 Chapter Ten

彩色钻石
Colored Diamonds

一、彩色钻石的基本特征
Basic Characteristics of Colored Diamonds

彩色钻石，你可能听到他们被称为"彩色钻石（Fancy Color Diamonds）"或"彩钻（fancies）"，这相对白色钻石而来的。钻石本来就有很多种颜色，白色钻石是最常见也是最便宜的一种，彩色钻石不常见，价格也更高。它的颜色包括黄色、褐色、红色、橙色、蓝色以及绿色等，都被称为"彩色钻石"。

For colored diamonds, you may hear them referred to as "Fancy Color diamonds" or "fancies", which is in contrast to white diamonds. Diamonds come in many colors, white diamonds are the most common and cheapest, colored diamonds are less common and more expensive. Its colors include yellow, brown, red, orange, blue and green, and are called "colored diamonds".

粉钻吊坠

黄钻戒指

绿钻戒指

小雨老师矿区宝石交易

小雨老师矿区宝石交易

小雨老师矿区收购宝石

Basic Characteristics of Colored Diamonds

Three points need to be paid attention to when classifying the color grade of colored diamonds: color, saturation and hue.

二、如何鉴赏和投资收藏彩色钻石
How to Appreciate and Invest in the Colored Diamond Collection

　　我们主要从以下6个维度去鉴赏和投资彩色钻石，我在每一个鉴赏指标分析里都给出了目前的市场情况以及相应指标的价格影响，也参考了一些拍卖价格，大家可以参考。

　　We mainly appreciate and invest colored diamonds from the following six aspects. In each appreciation index analysis, I have given the current market situation and the price impact of the corresponding indicators, and also referred to some auction prices for your reference.

（一）彩色钻石的鉴赏
Appreciation of Colored Diamonds

1.颜色
Color

是指彩色钻石有哪些颜色。
It refers to the colors of colored diamonds.

颜色
Color

彩钻颜色等级划分需注意三点：颜色、饱和度、色调。彩色钻石的颜色范围非常广泛，包括黄色、粉红色、蓝色、绿色、橙色、红色等。这些颜色的产生是由于钻石中微量的杂质元素或晶格缺陷所形成的。彩色钻石的分级并不是像白色钻石那样延用字母排序的方式，而是按照颜色的深浅即饱和度来分的，颜色分级从低到高依次为faint(微弱的)、very light(非常浅的)、light(浅的)、fancy light(较浅的)、fancy(正常的)、fancy dark(暗的)、fancy intence(较深的)、fancy deep(深的)、fancy vivid(鲜艳的)。以粉色为例，分为fait pink(微粉)，light pink(淡粉色)，fancy light pink(彩淡粉色)，fancy pink(彩粉色)，fancy dark pink(暗彩粉色)，fancy intense pink(浓彩粉色)，fancy deep pink(深彩粉色)，fancy vivid pink(艳彩粉色)。

Color

The classification of color diamonds should pay attention to three points: color, saturation, and tone. The color range of colored diamonds is very wide, including yellow, pink, blue, green, orange, red and so on. These colors are produced due to trace amounts of impurity elements or lattice defects in the diamond. Colored diamonds are not graded alphabetically like white diamonds, but according to the depth of color, or saturation, The color scale from low to high is faint, very light, light, fancy light, fancy, fancy dark, fancy intense, fancy deep, fancy vivid. Take pink as an example: fait pink, light pink, fancy light pink, fancy pink, fancy dark pink, fancy intense pink, fancy deep pink, fancy vivid pink.

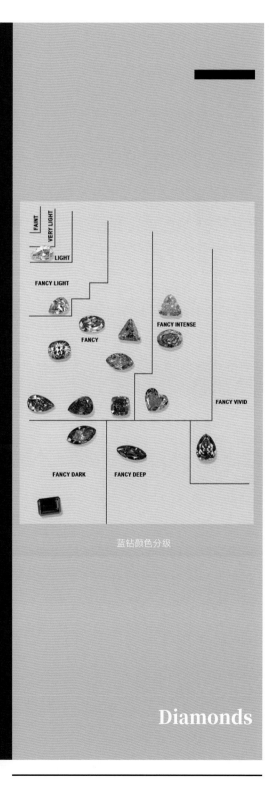

蓝钻颜色分级

Colored Diamonds

红色

Red 是彩色钻石里面价值最高的颜色，也是最稀有的颜色，至今世界上已知的纯红色钻石不足50颗。颜色从淡到浓的红色都有，纯红的红钻非常稀有。它是因为内部碳原子产生缺失形成的，你可以理解成它发生了基因突变，所以形成了红色，但是发生这样的基因突变，概率是很小的，红色钻石大约只占所有彩钻的0.01%，可见其稀有程度。目前主要产自澳大利亚的阿盖尔矿区。重5.11克拉的钻石穆萨耶夫红钻（Moussaieff Red），是世界上最大的红钻，呈三棱形切割，内部无瑕。早在2003年该钻报价已超700万美元。它于20世纪中期在巴西被一位农民发现，最初原石重量13.9克拉。这颗红宝石色的钻石最初被命名为"红盾"，但在穆萨耶夫珠宝公司收购之后，其名称被改为"穆萨耶夫红钻"。

Red

It is the most valuable color in colored diamonds, and it is also the rarest color. so far there are less than 50 pure red diamonds known in the world. The color can range from light to strong red, and pure red diamonds are very rare. It is formed because of the loss of internal carbon atoms, you can understand that it has a genetic mutation causing the formation of red, but the probability of such a genetic mutation is very small. Red diamonds only account for about 0.01% of all color diamonds, which shows its rarity. It is mainly produced in the Argyle mines of Australia. The Moussaieff Red, a 5.11-carat diamond, is the largest red diamond in the world, with a triangular cut and a flawless interior. As early as 2003, the diamond was quoted for more than $7 million. It was discovered by a farmer in Brazil in the mid-20th century and originally weighed 13.9 carats. The ruby-colored diamond was originally named "Red Shield", but after its acquisition by Musayev Jewelry Company, the name was changed to "Musayev Red Diamond".

穆萨耶夫红钻

30克拉黄钻

黄色

Yellow 是彩色钻石里面最常见的颜色，也是彩色钻石入门款。它是氮元素致色的，相对于其他彩色钻石形成的复杂性，它比较简单，所以比较常见，价格也比较低。然而在整个天然钻石界，也只占0.1%，依然很稀有。

Yellow

It is the most common color in colored diamonds, and it is also the introductory color diamond. It is nitrogen color, relative to the complexity of the formation of other colored diamonds, it is relatively simple, so it is more common, and the price is relatively low. However, in the total natural diamond world, it accounts only for 0.1%, is still very rare.

绿色

Green 绿色是由于受到自然辐射从而晶格结构被改变造成的。"德莱斯顿（Dresden）"绿钻重40.8克拉，是世界上已知最大的绿色钻石。绿钻是彩钻家族中备受推崇的一种，代表耐力和稳定性，绿钻配饰总能展现出低调、沉稳、有亲和力的美。德莱斯顿绿钻，是一颗蓝绿色重41克拉的梨形钻石。因其藏于德国的德莱斯顿宝库中，故而得名"德莱斯顿绿"。

Green

The green color is caused by natural radiation that changes the lattice structure. The Dresden green diamond weighs 40.8 carats and is the largest known green diamond in the world. Green diamond is a highly respected color diamond family, representing endurance and stability. Green diamond accessories can always show the low-key, calm, friendly beauty. The Dresden Green Diamond is a blue-green 41 carat pear-shaped diamond. Because it was stored in the Dresden Treasury in Germany, it was named "Dresden Green".

德勒斯顿绿

蓝色

Blue 蓝色是因为有微量元素硼的介入。世界上只有20万颗钻石被认证蓝钻，而深蓝色钻石非常罕见，价格仅次于红钻，拍卖场上出现的蓝钻的颜色级别主要以fancy vivid blue 和fancy intense blue为主。戴比尔斯库利南浩宇之蓝（The De Beers Blue）重达15.10克拉，估值约4800万美元（折合人民币约3亿元）。

Blue

The blue color is due to the involvement of the trace element boron. Only 200,000 diamonds in the world are certified blue diamonds, and dark blue diamonds are very rare and the price is second only to red diamonds, and the color levels of blue diamonds appearing at auction are mainly fancy vivid blue and fancy intense blue. The De Beers Blue weighs 15.10 carats, valued at about $48 million (equivalent to about 300 million yuan).

戴比尔斯库利南浩宇之蓝

粉色
Pink

粉色是由于在钻石产生过程中晶格结构发生了扭曲而成的。粉钻非常稀有，在自然界中发现的粉钻仅有10%的重量超过0.2克拉。产量非常稀少，现今90%以上的粉钻来自澳大利亚的阿盖尔矿区。一般来说，紫粉的价值是最高的，其次是纯粉，再次是橙粉（又叫橘粉）、棕粉。粉红之星（Pink Star），原石重132.5克拉，其颜色及净度皆获美国宝石学院评定粉红钻的最高评级。

Pink

The pink color is caused by the distortion of the lattice structure during the production of diamonds. Pink diamonds are extremely rare, with only 10% of those found in nature weighing more than 0.2 carats. Its production is very rare, with more than 90% of pink diamonds today coming from the Argyll mines in Australia. In general, the value of purple powder is the highest, followed by pure pink, orange pink (also known as orange pink), brown pink. The Pink Star, which weighs 132.5 carats in raw stone, has the highest rating for pink diamond color and clarity from the Gemological Institute of America.

粉红之星

紫色
Purple

紫色是由于氢元素导致的。目前世界上几乎没有纯紫色的钻石，即使有也仅仅是很小的颗粒。全球总产量不到100克拉，拥有神秘的气质。目前只有两颗相对知名的紫钻，分别为皇家紫心紫钻（Royal Purple Heart Diamond）和至尊紫心勋章（Supreme Purple Heart）。

Purple

The purple color is caused by hydrogen. At present, there are almost no pure purple diamonds in the world, and if there are, they are only very small particles. With a global output of less than 100 carats, it has a mysterious temperament. There are only two relatively well-known purple diamonds, the Royal Purple Heart Diamond and the Supreme Purple Heart.

皇家紫心紫钻

橙色
Orange

橙色钻石与黄色钻石一样，颜色的形成是受到内部碳晶体结构和氮元素的影响。宝石学家们认为橙色钻石是彩色钻石中继红色、蓝色、粉色、绿色或紫色之后的第五种最稀有的颜色。一颗3.22克拉的艳彩橙钻，净度为SI2，2016年12月5日，北京匡时拍卖行拍卖一颗温莎夫人（Lady Windsor）黄钻，起拍价达到30,000,000-35,000,000元人民币。

Orange

Orange diamonds, like yellow diamonds, are influenced by the internal carbon crystal structure and nitrogen. Gemologists consider orange diamonds to be the fifth rarest color of colored diamonds after red, blue, pink, green, or purple. A 3.22 carat rich orange diamond with clarity SI2. On December 5, 2016, Beijing Kuangshi Auction House auctioned a Lady Windsor yellow diamond with a starting price of 30,000,000 to 35,000,000 yuan.

温莎夫人黄钻

黑色
Black

大多数天然黑色钻石的颜色来自大量或微小的矿物内含物一如石墨、黄铁矿或赤铁矿等。黑色钻石的市场曾经是很低迷的，但现在人们对黑色宝石的需求日益增加。

Black

Most natural black diamonds get their color from large or small mineral inclusions such as graphite, pyrite or hematite. The market for black diamonds used to be very depressed, but now the demand for black gems is increasing.

555.55ct 黑色钻石裸石

Colored Diamonds

2.净度
Clarity

就是钻石内部的干净程度，彩色钻石的净度标准跟白色钻石是一样的。
It is the cleanliness of the inside of the diamond, the clarity standard of colored diamonds is the same as that of white diamonds.

1　无暇级　Flawless

无瑕级 (FL) 在 10 倍放大镜下观察，钻石没有任何内含物或表面特征。
Flawless grade (FL) When viewed under a 10x magnifying glass, the diamond has no inclusions or surface features.

2　内无瑕级　Internally Flawless

内无瑕级 (IF) 在 10 倍放大镜下观察，无可见内含物。
The internal flawless grade (IF) When viewed under a 10x magnifying glass and no inclusions are visible.

3　极轻微内含级　Very Slightly Included Grades

极轻微内含级（VVS1 和 VVS2）在 10 倍放大镜下观察，钻石内部有极微小的内含物，即使是专业鉴定师也很难看到。
Very slight inclusions grade (VVS1 and VVS2) When viewed under a 10x magnifying glass, diamonds have extremely tiny inclusions that are difficult for even professional examiners to see.

4　轻微内含级　Slightly Included Grades

轻微内含级（VS1 和 VS 2）在 10 倍放大镜下观察，钻石的内部可以看到微小的内含物。
The micro-inclusion grades (SI1 and SI2) Visible inclusions when viewed under a 10x magnifying glass.

5　微内含级　Slightly Included Grades

微内含级（SI1 和 SI2）在 10 倍放大镜下观察，钻石有可见的内含物。
Slightly included grades (SI1 and SI2) have visible inclusions when viewed under 10x magnification.

6　内含级　Inclusions in Included Grade

内含级（I1、I2 和 I3）钻石的内含物在 10 倍放大镜下明显可见，并且可能会影响钻石的透明度和亮泽度。
Inclusions grades (I1, I2 and I3) Inclusions are clearly visible under a 10x magnifying glass and may affect the clarity and brightness of the diamond.

钻石的内含物和表面特征通常都非常微小，只有专业的钻石鉴定师才能看到。肉眼看上去，VS1级钻石和SI2级钻石可能完全一样，但其整体品质却相差极大。因此，专业的钻石净度鉴定非常重要。最好是买SI1或以上的净度级别的，这样肉眼就看不到明显的瑕疵，净度等级越高越好。

The inclusions and surface features of diamonds are usually so tiny that only professional diamond examiners can see them. To the naked eye, a VS1 diamond and a SI2 diamond may look exactly the same, but their overall quality is very different. Therefore, professional diamond clarity identification is very important. It is best to buy a clarity level of SI1 or above, so that the naked eye can not see obvious defects, the higher the clarity level, the better.

钻石净度分级

3.切工
Cut

通常彩钻的切工都异形切割，除了圆形钻石外的切割都叫异形钻。彩钻不追求钻石的火彩，会根据彩钻原石的形状设计切磨能够最大保住重量的形状，最大地体现彩钻的价值，比如枕形、水滴形、心形等。所以彩钻如果是异形钻，他的GIA切工的分级，只包含抛光级别Polish跟对称性级别Symmetry，等级也是跟白钻一样包含EX（Excellent）、VG（Very good）、G（goog）、F（faint）、P（poor）。异形切割在抛光级别做到VG级别、对称性上面做到G切工或以上都是可以接受的。

Usually the cut of colored diamonds are shaped cutting, in addition to the round diamond cutting are called shaped diamonds. Color diamonds do not pursue the fire color of diamonds, will be based on the shape of the color diamond rough stone design cutting grinding can maximize the weight of the shape, the most reflect the value of color diamonds, such as pillow shape, water drop shape, heart and so on. Therefore, if the color diamond is a special-shaped diamond, his GIA cut classification only includes the Polish level and the Symmetry level, and the grade is also the same as the white diamond, including EX (Excellent), VG (Very good), G (good), F (faint), P (poor). It is acceptable for profiled cutting to achieve VG level in polishing level and G cut or above in symmetry level.

4.重量
Weight

彩色钻石相比白色钻石而言比较稀有，所以在重量上我们选择1克拉的高品质的彩色钻石也比较保值，不需要刻意追求大小。打个比方，粉钻里面的阿盖尔粉钻，1克拉的高品质粉钻是要过百万人民币的，1克拉的红钻更加贵重。因为彩色钻石更注重颜色，所以说在颜色好的情况下，克拉重量小一点也比较保值。

Colored diamonds are relatively rare compared to white diamonds, so in terms of weight, we choose 1 carat high-quality colored diamonds are also relatively stable, and there is no need to deliberately pursue size. For example, the Argyle pink diamond inside the pink diamond, the high-quality 1 carat pink diamond is more than one million yuan, and the 1 carat red diamond is more valuable. Because colored diamonds pay more attention to color, it is said that in the case of good color, the carat weight is smaller will also be better value preservation.

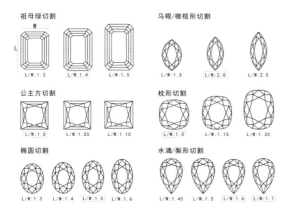

异形宝石切割

16+克拉黄钻戒指

5.彩钻的荧光级别
The Fluorescence Level of Color Diamonds

彩钻的荧光级别跟白钻一样，GIA根据钻石荧光的强弱程度分为N(无)、F(弱)、M(中等)、S(强)、VS(很强)，这几个级别。

The fluorescence level of color diamonds is the same as that of white diamonds, and GIA is divided into N(no), F(feeble), M(medium), S(strong), and VS (very strong) according to the intensity of diamond fluorescence.

钻石荧光

6.证书
Certificate

GIA目前是钻石这个领域比较权威的证书。

GIA is currently the most authoritative certificate in the field of diamonds.

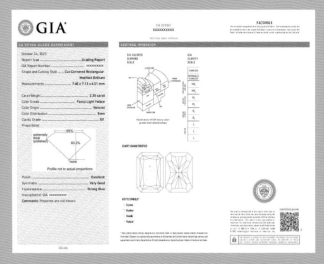

GIA彩色钻石证书

小雨老师 经验分享
Experience Sharing from Ms. Xiaoyu

我想跟大家解开一个误会，很多人觉得白色钻石现在都已经有人工培育钻石了，所以白色钻石价格下跌。但是我想说的是，白色钻石价格跌不是因为人工培育的白色钻石出现，而是白色钻石本身在地球上的产量就很大，这才是它价格下降的原因。那很多人又问，如果也有人工培育彩色钻石出现的话，那是不是彩色钻石价格也会下跌？其实彩色钻石早就已经有了人工合成的，在河南就有人工培育彩色钻石，但是并没有影响到天然的彩色钻石的价格。因为我们的宝石投资的逻辑是这个宝石本身就稀有且漂亮，跟它能不能人工培育没有关系。彩色钻石本身就稀有，而白色钻石自然界中本身就多，所以白色钻石相对收藏价值低，逻辑是这样的才对。

I would like to clear up a misunderstanding with you. Many people think that white diamonds have now been artificially cultivated diamonds, so the price of white diamonds has fallen. But what I want to say is that the price of white diamonds has fallen not because of the emergence of artificially cultivated white diamonds, but the production of white diamonds themselves on the earth is large, which is the reason for its price decline. Then many people may ask, if there is also artificial cultivation of colored diamonds, will the price of colored diamonds also fall? In fact, colored diamonds have long been synthetic, there are artificial cultivation of colored diamonds in Henan, but it has not affected the price of natural colored diamonds. Because the logic of our gemstone investment is that the gemstone itself is rare and beautiful, regardless of whether it can be cultivated or not. Colored diamonds themselves are rare, and white diamonds are abundant in nature, so white diamonds are not valuable. This is the right logic.

Colored diamonds

(二) 彩色钻石的投资收藏建议
Colored Diamond Investment and Collection Suggestions

1.彩色钻石拍卖纪录
Colored Diamonds Auction Records

时间 Time	拍卖行 Auction House	拍品 Item	成交价 Closing Price
2023年5月 May, 2023	苏富比香港 Sotheby's Hong Kong	1.06克拉彩红色钻石配钻石戒指 1.06 carat colored red diamond with diamond ring	成交价 4,410,000 港元 Sale Price HKD 4,410,000
2024年1月 January, 2024	苏富比香港 Sotheby's Hong Kong	宝格丽14.5克拉浓彩黄色钻石戒指 Bulgari 14.5 carat rich yellow diamond ring	成交价 2,540,000 港元 Sale Price HKD 2,540,000
2023年10月 October, 2023	苏富比香港 Sotheby's Hong Kong	无际之蓝11.28克拉浓彩蓝色钻石戒指 Endless blue 11.28 carat diamond ring in blue color	成交价 198,220,000 港元 Sale Price HKD 198,220,000
2023年11月 November, 2023	苏富比日内瓦 Sotheby's Geneva	4.42克拉浓彩绿色钻石戒指 4.42 carat diamond ring in rich green color	成交价 1,633,000法郎 Sale Price CHF 1,633,000
2015年11月 November,2015	佳得士日内瓦 Christie's Geneva	16.08克拉浓彩粉色钻石戒指 16.08 carat rich pink diamond ring	成交价 28,725,000法郎 Sale Price CHF 28,725,000

2.彩色钻石涨价的原因
Reasons for the Price Increase of Colored Diamonds

彩色钻石的形成跟白色钻石不一样,彩色钻石是因为致色元素或者原子变异形成的颜色,而且颜色还要均匀,保证4C都比较好的情况下,才能形成一颗高品质的彩色钻石,所以彩色钻石比白色钻石稀有很多。以粉色钻石为例,阿盖尔矿区出产市面上90%的粉钻,但是2020年已经宣布封矿,即便发现新的矿区,也需要15年以上的开采时间才能面向消费者。对于早期投资粉钻的人来说是好事,资产会再次增值,但是对于想买还没有买粉钻的人来讲,相当于又涨价了一个阶段。所以只要物色到高品质的宝石,价格合适一定要趁早投资。

The formation of color diamonds is not the same as white diamonds. Color diamonds are formed because of color elements or atomic variation, and the color should be even, to ensure that the 4C is better, so as to form a high-quality color diamond, so color diamonds are much rarer than white diamonds. In the case of pink diamonds, the Argyll mining area produces 90% of the pink diamonds on the market, but it has been announced that the mine has been closed in 2020, and even if a new mining area is found, it will take more than 15 years to mine before it is ready for consumers. It is good for those who invested in pink diamonds early, and the assets will appreciate again, but for those who want to buy pink diamonds, it is equivalent to another stage of price increase. So as long as you find high-quality gems, the price is right, you must invest as early as possible.

3.彩色钻石投资案例
Colored Diamond Investment Case

2019年,我还在上学的时候,一个开美容院的姐姐来学校找我,让我帮她物色一颗高品质的粉钻,她想结婚用。2019年9月我就去香港国际珠宝展帮她看看粉色钻石,当时是一个美国珠宝商,我讲了好久的价,1克拉fancy intence 颜色,最后成交价格是280万元人民币。2024年同样品质的粉色钻石从美国拿货的批发价格都是400万元人民币。

In 2019, when I was still in school, a sister who ran a beauty salon came to my school and asked me to help her find a high-quality pink diamond that she wanted to use for her wedding. In September 2019, I went to the Hong Kong International Jewelry Fair to help her look at the pink diamond. There was an American jeweler at that time, I talked about the price for a long time, 1 carat fancy intense color, and the final closing price was 2.8 million yuan. In 2024, the wholesale price of pink diamonds of the same quality from the United States is 4 million yuan.

4.彩色钻石的投资收藏建议
Suggestions for Investment Collection of Colored Diamonds

(1)权威证书:GIA证书。
Authoritative Certificate: GIA certificate

(2)高品质彩色钻石:优先选择1克拉以上,颜色达到fancy intence、净度达到SI以上的。
High quality color diamonds: preferred to choose more than 1 carat, color to achieve fancy intense, clarity of SI or above.

(3)市场趋势:关注国际拍卖价格以及国际珠宝展的价格,也可以咨询名贵珠宝投资顾问。
Market Trends: Pay attention to international auction prices and international jewelry show prices, you can also consult luxury jewelry investment advisers.

小雨老师 经验分享
Experience Sharing from Ms. Xiaoyu

彩色钻石的投资是比彩色宝石金额大很多的,大多数1克拉的高品质彩色钻石投资金额都是超过100万元人民币的,而1克拉的高品质彩色宝石,以高品质缅甸红宝石为例,价格只在25万元人民币左右。彩色钻石在国际上的流通度和知名度都非常好,投资收益是比较稳定的。我在这本书的下篇会讲到,名贵珠宝的投资分为投资型和投机型,彩色钻石以其稳定的涨幅,属于投资型。

The investment in colored diamonds is much larger than the amount of colored gems, most of the 1 carat high-quality colored diamonds investment amount is more than 1 million yuan, and 1 carat high-quality colored gems, taking high-quality Myanmar ruby as an example, the price is only about 250,000 yuan. The international circulation and visibility of colored diamonds are very good, and the investment income is relatively stable. As I will explain in the next part of this book, the investment of luxury jewelry is divided into investment type and speculation type, and colored diamonds are investment type with their steady increase.

Luxury Jewelry Appreciation and Investment

The Beauty of Conch Pearls

第十一章
11 Chapter Eleven

海螺珠
Conch Pearls

一、海螺珠的基本特征
Basic Characteristics of Conch Pearl

海螺珠也叫海螺珍珠（Conch Pearls）或者孔克珠（Melo Pearlo）。海螺珠的产量非常稀少，被誉为"珍珠中的爱马仕"。据了解，平均每一万只大凤螺中，只有约一颗可用的海螺珠，而且只有约一成符合制成珠宝首饰的标准与质量。因此，全球每年可用于制成珠宝的海螺珍珠数量非常有限，少于900颗。所以海螺珠过去都是皇室所有，普通人连拥有的渠道都没有。海螺珠也是拍卖场上的宠儿，很多珠宝收藏家或者珠宝投资者都希望收藏一颗海螺珠。近几年海螺珠在拍卖会上风头正盛，让很多朋友误以为海螺珠是一种突然出现的贵重珠宝。但其实早在19世纪的维多利亚时期，就已经出现许多精美的海螺珠作品。纵观不同设计风格，从维多利亚时期、新艺术时期到装饰艺术风格时期，海螺珠从未缺席。目前国内珠宝收藏者也越来越关注海螺珠，未来还有很多的涨价空间。

Conch Pearls are also called Melo Pearlo. The production of conch pearl very rare, known as the "Hermes of pearls". It is understood that on average, there is only about one available conch pearl in every 10,000 large conchs, and only about 10% meet the standards and quality of jewelry. As a result, the number of conch pearls that can be used to make jewelry worldwide is very limited, less than 900 per year. So conch pearl used to be owned by the royal family, and ordinary people didn't even have access to them. Conch pearls are also the beloved one of the auctions, and many jewelry collectors or jewelry investors want to collect a conch pearl. In recent years, conch pearls in the auction limelight are prosperous, so that many friends mistakenly think that conch pearls are a sudden appearance of valuable jewelry. But in fact, as early as the Victorian period in the 19th century, there had been many exquisite conch pearls. Throughout the different design styles, from the Victorian period, Art Nouveau to the Art Deco period, conch pearls have never been absent. At present, domestic jewelry collectors are also increasingly concerned about conch pearls, and there is still a lot of room for price increases in the future.

小雨老师参与苏富比拍卖

海螺珠戒指

小雨老师日本考察

海螺珠项链

小雨老师日本考察

小雨老师日本考察

Basic Characteristics of Conch Pearl

Formation of Conch Pearls: Unlike pearls formed inside oysters, conch pearls originate from a pink queen conch residing in the Caribbean Sea.

Conch pearls

二、如何鉴赏和投资收藏海螺珠
How to Appreciate and Invest in the Conch Pearl Collection

我们主要从以下7个维度去鉴赏和投资一颗海螺珠，我在每一个鉴赏指标分析里都给出了目前的市场情况以及相应指标的价格影响，也参考了一些拍卖价格，大家可以参考。

We mainly appreciate and invest a conch from the following seven aspects. In each appreciation index analysis, I have given the current market situation and the price impact of the corresponding indicators, and also referred to some auction prices for your reference.

(一)海螺珠的鉴赏
Appreciation of Conch Pearls

1.产地
Origin

是指海螺珠的出产地，对价格影响挺大的。
It refers to the origin of conch pearls, which has a great impact on the price.

海螺珠的形成

与一般在牡蛎体内形成的珍珠不同，海螺珠产自一种在加勒比海居住的粉红色大凤螺体内。这种海螺本身就十分漂亮，像号角海螺，颜色粉中带橙。据说每1000到2000个女王凤凰螺中才会发现一颗海螺珠，其中只有10%可以达到宝石级的品质。因此海螺珠是最珍贵的天然珍珠之一。海螺珠的形成也与珍珠不同，海螺珠不能人工养殖，当有刺激性的物体比如断裂的贝壳或其他蠕虫进入体内后，它自身分泌的钙质会结晶，从而形成海螺珠。

Unlike pearls, which are usually formed in oysters, conch pearls come from the body of a large pink conch that lives in the Caribbean Sea. The conch itself is very beautiful, like the horn conch, pink with orange. It is said that only one conch pearl can be found in every 1,000 to 2,000 Queen Phoenixes, of which only 10% can achieve gem quality. Therefore, conch pearls are one of the most precious natural pearls. The formation of conch pearls is also different from pearls, conch pearls can not be cultured, when irritating objects such as broken shells or other worms into the body, its own secretion of calcium will crystallize, thus forming conch pearls.

加勒比海螺珠

加勒比海
The Caribbean Sea

加勒比海是最著名的海螺珠产地，这里阳光温暖、海水清澈，非常适合大凤螺栖息。围绕在加勒比海西海岸的多米尼加共和国、洪都拉斯、牙买加等地都是大凤螺的栖息地。但是由于孕育海螺珠的大凤螺不仅仅造就了绝美的珍珠，更是难得的美味，如今大范围的猎杀使得除了3个海螺产出国以外的所有国家都订立了严格的管制方案以保护大凤螺，这意味着市面上将会越来越少地出现天然的海螺珠，这也是海螺珠稀少的另一原因。海螺珠的形成完全归功于大自然，完全没有任何的人类的干预，这是海螺珠昂贵的原因之一。再加上发现海螺珠的概率大约是万分之一，在这些珍珠里能达到宝石质量的不到10%，因此让海螺珠倍显珍贵。

The Caribbean Sea The Caribbean is the most famous source of conch pearls, where the sun is warm, the water is clear, very suitable for the large conch habitat. The Dominican Republic, Honduras, and Jamaica, which surround the west coast of the Caribbean Sea, are the habitats of the giant phoenix snail. However, because the large conch that breeds conch pearls not only makes a beautiful pearl, but also a rare delicacy, today's large-scale hunting has made all countries except the three conch producing countries have set up strict control schemes to protect the large conch. This means that there will be less and less natural conch pearls on the market, which is another reason for the scarcity of conch pearls. The formation of conch pearls is entirely due to nature, completely without any human intervention, which is one of the reasons why conch pearls are expensive. In addition, the probability of finding conch pearls is about one in 10,000, and less than 10% of the gem quality can be reached in these pearls, so that conch pearls are more precious.

2. 大小
Size

大多数海螺珠都非常小,只有米粒大小,首饰中经常用到的是1到3克拉的海螺珠,大于5克拉就非常少见了。

图3.1 不同大小的海螺珠

Most conch pearls are very small, only the size of a grain of rice, and 1 to 3 carats are often used in jewelry, but larger than 5 carats are very rare.

3. 形状
Shape

常见的是近椭圆形、水滴形、近圆形以及不规则形状。天然珍珠的形状通常都不规则,对于这一点不必太过苛刻,毕竟每一颗天然珍珠都弥足珍贵且独一无二。

The common ones are near-oval, drop-shaped, near-round, and irregular shapes. The shape of natural pearls is usually irregular, and it is not necessary to be too harsh on this point, after all, each natural pearl is precious and unique.

不同形状的海螺珠

4. 颜色
Color

它是天然珍珠中唯一的粉色珍珠,它的颜色高贵而浪漫。人们也赋予了它很多美丽的名字,其中最受欢迎的是火烈鸟粉(Flamingo Pink)、樱花粉(Cherry Blossom Pink)和三文鱼粉(Salmon Pink)。一般是浅粉色到深粉色,常见的还有橙粉色、棕粉色、黄色和红色。其中粉色和橙粉色是最受欢迎的颜色。海螺珠的颜色与女王凤凰螺的大小和成熟度有关。女王凤凰螺3到4年才能成熟,而它的平均寿命是20到30年,一般海螺生长的年龄越长越成熟,内部的颜色越红,才有可能形成宝石级的海螺珠。我们依据饱和度将海螺珠的颜色分为五个等级:Strong、vivid、intense、light、fair。

It is the only pink pearl among natural pearls, and its color is noble and romantic. People have also given it many beautiful names, among which the most popular are Flamingo Pink, Cherry Blossom Pink and Salmon Pink. They are usually light pink to dark pink, with orange pink, brown pink, yellow and red being common. Pink and orange pink are the most popular colors. The color of the conch pearls is related to the size and maturity of the Queen conch. The Queen conch takes 3 to 4 years to become mature, and its average life span is 20 to 30 years. Generally, the older the conch growing, the more mature it become, the more red the internal color, and it will be more possible to form gem-grade conch pearls. We rated the color of the conch pearls into five levels based on saturation :strong, vivid, intense, light, and fair.

Vivid评级海螺珠

5.净度
Clarity

◆

天然海螺珠的表面有时会有白色的斑点，经常在珍珠的两端出现，行业内称为"钙点"。当钙点少量出现在珍珠两端时，对外观影响不大，但是当钙点在珍珠表面大面积出现时，就会影响海螺珠的美观了。根据海螺珠表面钙点的多少分为五个等级:clean、very lightly、spottedlightly、spotted moderately、spotted heavily spotted。

The surface of natural conch pearls sometimes has white spots, often appearing at both ends of the pearl, known in the industry as "calcium spots." When a small amount of calcium appears at both ends of the pearl, it has little effect on the appearance, but when the calcium point appears in a large area on the surface of the pearl, it will affect the beauty of the conch. According to the number of calcium points on the surface, the conch pearls are divided into five grades: clean, very lightly, spotted lightly, spotted moderately, and spotted heavily spotted.

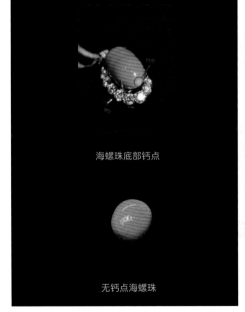

海螺珠底部钙点

无钙点海螺珠

海螺珠 Conch Pearls

6.特征
Features

◆

海螺珠的火焰纹：海螺珠和天然珍珠相同，有着一层层的同心圆的珠层结构，化学成分同是方解石，但是海螺珠没有像珍珠那样的珍珠层。这和方解石的具体结构有关，就好像盖了两座房子，垒砖的方式不同，天然珍珠是一层一层垒在一起，而海螺珠的方解石结构则是纵横交错的，这种随意的结构恰恰形成了海螺珠独特的火焰纹结构。在阳光下转动海螺珠，有火焰般的纹理伴随丝绢光泽，如流动的异彩，展现出一种独有的细腻和柔美。最好的海螺珠火焰纹是在整个珍珠表面都可以清晰地见到，但这样的珍珠非常稀有。excellent-very good的级别时，会在证书上的Special Comment(特别名称)一栏写:fire flame，代表火焰纹非常好。根据海螺珠表面火焰纹是否肉眼可见以及覆盖珍珠表面的多少，火焰纹分为五个等级：excellent、very good、good、fair、no flame。

Conch pearls flame pattern: conch pearls and natural pearls are the same, with layers of concentric circle pearl structure, the chemical composition is the same calcite, but conch pearls do not have a pearl layer like pearls. This is related to the specific structure of calcite, just like building two houses, the way of laying bricks is different, the natural pearl is layered together, and the calcite structure of conch pearls is crisscross. This random structure is exactly the formation of conch pearls unique flame pattern structure. Turning conch pearls under the sun, there are flame-like textures accompanied by silky luster, such as flowing colors, showing a unique delicate and soft beauty. The best conch pearl flame pattern is clearly visible throughout the pearl surface, but such pearls are very rare. An excellent-very good rating will indicate "fire flame" in the Special Comment column on the certificate, which means the flame print is very good. Based on whether the flame pattern on the surface of the conch pearl is visible to the naked eye and how much it covers the surface of the pearl, the flame pattern is divided into five grades: excellent, very good, good, fair, and no flame.

Fire Flame评级海螺珠

7.证书
Certificate

不同的证书代表不同的含金量。
Different certificates represent different gold content.

01
GUILD

目前吉尔德证书对海螺珠的分级做得较完善。
At present, GUILD certificate on the classification of conch pearls has done more perfect.

吉尔德证书

(二)海螺珠的投资收藏建议
Conch Pearl Investment and Collection Suggestions

1. 海螺珠的拍卖纪录
An Auction Record for Conch Pearls

时间 Time	拍卖行 Auction House	拍品 Item	成交价 Closing Price
2023年4月 April, 2023	苏富比香港 Sotheby's Hong Kong	海螺珠配粉红色刚玉及钻石项链 Conch pearls with pink corundum and diamond necklace Size: 15.13 mm X 11.88 mm X 10.43 mm	成交价 863,600 港元 Sale Price HKD 863,600
2022年5月 May, 2022	佳士得香港 Christie's Hong Kong	海螺珍珠 钻石和红宝石戒指 十颗不同形状的海螺珍珠 直径约8.02至5.70毫米 Conch Pearls, Diamond and Ruby rings. Ten conch pearls of different shapes, about 8.02 to 5.70 mm in diameter	成交价 151,200 港元 Sale Price HKD 151,200
2019年11月 November, 2019	佳士得香港 Christie's Hong Kong	海螺珍珠野生珍珠和钻石戒指 Conch pearls, wild pearls and diamond rings	成交价 150,000 港元 Sale Price HKD 150,000
2019年10月 October, 2019	苏富比香港 Sotheby's Hong Kong	海螺、珍珠和钻石戒指 尺寸：14.10mmX13.66mmX10.00mm Conch, pearl and diamond rings Size: 14.10 mm X 13.66 mm X 10.00 mm	成交价 562,500 港元 Sale Price HKD 562,500

2. 海螺珠投资收藏建议
Investment and Collection Advice for Conch Pearls

从投资收藏的角度：我觉得收藏一颗海螺珠首先要从以下5个方面去评估价值。第一，一定要有非常明显的火焰纹，因为这是它典型的特征，火焰纹越明显，价值越高。第二，颜色一定要相对浓艳，不论是粉色、红色、黄色，都是颜色越浓郁，价格越高。第三，瑕疵度，尽量选择台面没有瑕疵的，两边有钙化点没关系的，因为在镶嵌的时候会遮住。第四，形状，整个形态尽量圆润。第五，大小，其他几个方面都特别好的话，即便小一点也是有价值的。在自己预算充足的情况下，我们肯定要选择大克拉的。但前提是，要能找到这样的海螺珠，高品质的海螺珠真的太稀有了。

From the perspective of investment collection: I think collecting a conch should firstly to evaluate the value from the following five aspects. First, there must be a very obvious flame pattern, because this is its typical characteristics, the more obvious the flame pattern, the higher the value. Second, the color must be relatively bright, whether it is pink, red, yellow, the more intense the color, the higher the price. Third, the degree of defects, try to choose the one whose mesa are without defects, and there are calcification points on both sides of the matter, because it will be covered when doing the mosaic. Fourth, the shape, the whole shape should be as round as possible. Fifth, size, if several other aspects are particularly good, even a small is also valuable. Given our budget, we definitely want to go with the big carat. But only if you can find such conch pearls, as high-quality conch pearls are really too rare.

小雨老师 经验分享
Experience Sharing from Ms. Xiaoyu

给大家讲一个小故事，2016年我跟老师一起去日本首饰学院进修，参观了三重县的akoya珍珠博物馆，大家鉴赏的都是akoya的珍珠。当我们要下课的时候，馆长说，我们现在要拿一个镇店之宝给大家看一看，我跟同学们都很期待，以为肯定会拿一颗很大的akoya的珍珠给我们长长眼。没想到他拿上来一颗非常非常漂亮的海螺珠，我当时都被惊艳到了，那种浓艳的粉色，火焰纹又非常明显，又特别大颗，至少10克拉，一颗特别完美的海螺珠，真的是稀世珍宝啊，确实算得上是镇馆之宝。

Let me tell you a short story. In 2016, I went to Japan Jewelry Institute with my teacher to study and visited the akoya Pearl Museum in Mie Prefecture. What we appreciated were akoya pearls. When we were about to finish class, the curator said, we are now going to take a top-one treasure to show you. My classmates and I were looking forward to it, thinking that they will definitely take a large akoya pearl to give us to see. We did not expect him to take up a very very beautiful conch, I was amazed at that time. The kind of rich pink, the flame pattern is very obvious, and particularly large, at least 10 carats. That was a particularly perfect conch, which really is a rare treasure, indeed can be regarded as the top- one treasure of the museum.

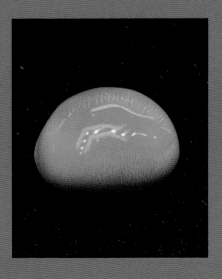

Luxury Jewelry Appreciation and Investment

Appreciate
Natural Pear

第十二章
12 Chapter Twelve

野生珍珠
Natural Pear

一、野生珍珠的基本特征
Basic Characteristics of Natural Pearls

野生珍珠英文名称为natural pear,又叫天然珍珠。曾经一颗叫做Marie Antoinette的天然野生珍珠在苏富比拍卖到了2.4亿元人民币的价格。之所以叫它野生珍珠或者天然珍珠,是因为它是在没有任何人为干预的情况下自然形成的,因此野生珍珠的形成是一种偶然而困难的过程。野生珍珠的蚌壳只能在清澈的水中存活,它们需要经历漫长的时间从小沙粒逐渐生长为一颗珍珠。如果水质不够清澈,珍珠在形成之前就会死亡。野生珍珠是完全天然生长的,因此其珍珠层非常厚,而珍珠层越厚,珍珠的光泽就越好。人工养殖的珍珠生长周期较短,珍珠层较薄,因此光泽不如野生珍珠强。

Wild pearl is also known as natural pearl. Once a natural wild pearl called Marie Antoinette was auctioned at Sotheby's for 240 million yuan. The reason why it is called wild pearl or natural pearl is that it is formed naturally without any human intervention, so the formation of wild pearl is an accidental and difficult process. The shell of a wild pearl can only survive in clear water, and it takes a long time for it to grow from a small grain of sand to a pearl. If the water is not clear enough, the pearl will die before it can form. Wild pearls grow completely naturally, so their nacre layer is very thick, and the thicker the nacre layer, the better the luster of the pearl. Cultured pearls have a shorter growth cycle and thinner nacre layer, so their luster is not as strong as that of wild pearls.

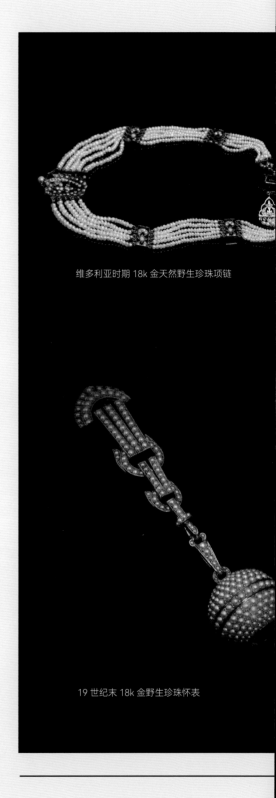

维多利亚时期 18k 金天然野生珍珠项链

19 世纪末 18k 金野生珍珠怀表

维多利亚时期 18k 金天然野生珍珠项链

  目前我们市面上99%的珍珠都是人工养殖珍珠，野生珍珠的形成是可遇不可求的，因此大多数市面上销售的天然珍珠都是古董级的珍品。因为在1916年之前，还没有人发明出养殖技术，所以在20世纪之前的珍珠都是天然珍珠。据历史记录，日本人在19世纪中叶开始研究和发展珍珠养殖技术。具体来说，日本研究人员小松次郎于1893年首次成功地培育出一颗珍珠，这标志着人工养殖珍珠的开始。随后，日本珍珠养殖技术得到不断改进和发展，成为世界上最重要的珍珠养殖国家之一。在养殖珍珠的过程中，日本人利用了贝壳内膜的特性，通过人工介入促使珍珠贝壳分泌珍珠层，并逐渐形成珍珠。这一技术的发明和创新对于珍珠行业的发展和推动起到了重要作用。

  At present, 99% of the pearls on the market are cultured pearls, and the formation of wild pearls is rare, so most of the natural pearls sold on the market are antique treasures. Because no one had invented farming techniques before 1916, pearls were natural until the 20th century. According to historical records, the Japanese began to research and develop pearl farming techniques in the mid-19th century. Specifically, Japanese researcher Jiro Komatsu successfully cultivated a pearl for the first time in 1893, which marked the beginning of cultured pearls. Subsequently, Japan's pearl culture technology has been continuously improved and developed, becoming one of the most important pearl culture countries in the world. In the process of cultivating pearls, the Japanese made use of the characteristics of the shell inner membrane, and promoted the pearl shell to secrete pearl layer through artificial intervention, and gradually formed pearls. The invention and innovation of this technology has played an important role in the development and promotion of the pearl industry.

  Compared to artificially cultivated pearls, wild pearls have a shorter growth cycle, thinner pearl layers, and consequently, their luster may not be as intense as that of wild pearls.

那如何区分野生和养殖珍珠呢？对肉眼来说基本上是不可能完成的任务，因为养殖珍珠的技术现在已经到了一个非常高的水平，它们的外观和品质都与野生珍珠相似。但是以下一些建议可以帮助你识别两者的差异。第一，外形和完整性。野生珍珠往往更不规则，因为它们在自然环境中形成，受到各种不可预测的环境因素的影响，而养殖珍珠是更规则的圆形或者半圆形。第二，表面的纹理，野生珍珠往往会有更多的瑕疵、凹陷或其他自然形成的表面特征。尽管养殖珍珠也有瑕疵，但由于在受控的环境中形成，其表面更加平滑，很有光泽。第三，光泽和颜色。虽然这不是一个可靠的区分方法，但野生珍珠往往具有更深沉和复杂的光泽，养殖珍珠的光泽可能更加均匀和明亮。第四，大多数的海水养殖珠都有一个坚硬的核心，但是野生珍珠整体由珍珠质组成，没有核心。

So how do you tell wild pearls from farmed pearls? It is basically an impossible task for the naked eye, as the technology of cultured pearls has now reached a very high level, and their appearance and quality are similar to those of wild pearls. But here are some tips to help you identify the differences. First, shape and integrity. Wild pearls tend to be more irregular because they form in the natural environment and are subject to a variety of unpredictable environmental factors, while cultured pearls are more regularly round or semi-circular. Second, the texture of the surface, wild pearls tend to have more blemishes, dents or other naturally occurring surface features. Although cultured pearls also have imperfections, their surface is smoother and shiny due to being formed in a controlled environment. Third, gloss and color. While this is not a reliable method of differentiation, wild pearls tend to have a deeper and more complex sheen, and farmed pearls may have a more even and brighter sheen. Fourth, most mariculture pearls have a hard core, but wild pearls as a whole are composed of pearl, and there is no core.

小雨老师经验分享
Experience Sharing from Ms. Xiaoyu

虽然野生珍珠很稀有，但不是所有野生珍珠价格都很高。只有大颗粒、光泽好、比较圆的野生珍珠才有价值。高品质的野生珍珠价格过亿，但是一些小颗粒的、形状不规则的野生珍珠只需要几千元人民币。

Although wild pearls are rare, not all wild pearls are highly valued. Only large grains, good luster, relatively round wild pearls are valuable. High-quality, wild pearls cost more than 100 million yuan, but some small particles, irregular shape of wild pearls only cost a few thousand yuan.

二、如何鉴赏和投资收藏野生珍珠
How to Appreciate and Invest in the Natural Pearl Collection

  我们主要从以下8个维度去鉴赏和投资一颗野生珍珠，我在每一个鉴赏指标分析里都给出了目前的市场情况以及相应指标的价格影响，也参考了一些拍卖价格，大家可以参考。

  We mainly appreciate and invest a wild pearl from the following eight aspects. In each appreciation index analysis, I have given the current market situation and the price impact of the corresponding indicators, and also referred to some auction prices for your reference.

(一) 野生珍珠的鉴赏
Appreciation of Natural Pearl

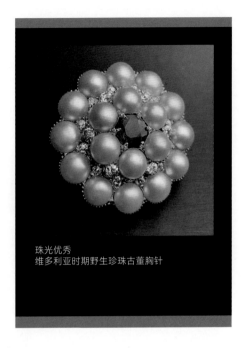

珠光优秀
维多利亚时期野生珍珠古董胸针

1. 珠光
Pearlescent

一定是珠光，珠光够好，其他都是可以不计的。

It must be pearlescent that comes first for evaluation. If pearlescent is good enough, then everything else is negligible.

2. 颜色
Color

每一颗珍珠都是由体色跟伴色以及光晕组成的，光晕伴色越丰富，它的珠光就更丰富，价值就越高。

Each pearl is composed of body color and companion color and halo, the richer the halo companion color, the richer its pearl, the higher the value.

3. 形状
Shape

肯定越圆的价值是越高的。

Surely, the rounder, the higher value.

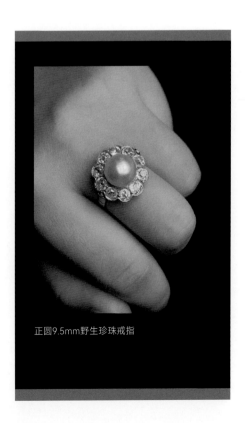

正圆9.5mm野生珍珠戒指

正圆野生珍珠　　　　野生珍珠

4. 大小
Size

大小也就是我们通常说的点位。珠子越大就意味着它在贝壳里面的时间越长，就越贵。

The size is what we usually call the point position. The larger the bead, the longer it stays inside the shell, the more expensive it will be.

5. 瑕疵度
Defect Degree

无瑕不成珠,我们尽量挑选瑕疵非常少的。

No flawless no pearl, but we try to choose very few defects.

6. 珠层厚度
Coating Thickness

涂层厚度是没有办法用肉眼看到的,要借助实验室的一些仪器。

The thickness of the coating can not be seen with the naked eye, it has to borrow help of some laboratory instruments.

7. 成套
Set

两颗珠子的价格不是 1+1>2,是可能成倍增长的。如果可以收集到一整套项链,那么价格可能会是每颗珠子价格之和的 10 倍。

The price of two pearls is not 1+1>2. It is possible to multiply an effect. If the entire necklace can be collected, the price may be 10 times the sum of each pearl.

野生珍珠胸针耳环套装

8. 证书
Certificate

高品质的野生珍珠最好出一个 SSEF 或者古柏林证书,证书上会显示这颗珍珠是野生珍珠。

High-quality wild pearls are best issued with an SSEF or Gubelin certificate, which will indicate that the pearl is a wild pearl.

SSEF 野生珍珠证书

GUBELIN野生珍珠证书

(二)野生珍珠的投资收藏建议
Natural Pearl Investment and Collection Suggestions

1.野生珍珠拍卖纪录
Wild Pearl Auction Record

时间 Time	拍卖行 Auction House	拍品 Item	成交价 Closing Price
2023年11月 November, 2023	佳士得香港 Christie's Hong Kong	卡地亚野生珍珠戒指 由65颗和65颗天然珍珠组成， 尺寸约为11.75至5.35毫米 Cartier Wild Pearl ring It consists of 65 and 65 natural pearls with a size of approximately 11.75 to 5.35 mm	成交价 6,930,000 港元 Sale Price HKD 6,930,000
2023年6月 June, 2023	苏富比纽约 Sotheby's New York	卡地亚野生珍珠戒指 珍珠尺寸：15.1mm×15.0mm Cartier wild pearl ring Pearl size: 15.1mm×15.0mm	成交价 139,700 美元 Sale Price USD 139,700
2020年11月 November, 2020	苏富比日内瓦 Sotheby's Hong Kong	宝格丽天然珍珠配钻石戒指 浅褐灰色天然珍珠，重35.21克拉， 尺寸为18.23mm×18.56mm×14.26mm Bulgari natural pearl with diamond ring Light brown grey natural pearl, 35.21 carats, size: 18.23 mm×18.56 mm×14.26 mm	成交价 157,500 法郎 Sale Price FR 157,500
2021年10月 October, 2021	苏富比香港 Sotheby's Hong Kong	天然珍珠配钻石及珍珠项链，卡地亚钻石链扣 由51颗和49颗天然珍珠组成， 尺寸约为10.30到5.00毫米 Natural pearls with diamond and pearl necklace, Cartier diamond chain link It consists of 51 and 49 natural pearls with a size of about 10.30 to 5.00 mm	成交价 2,016,000 港元 Sale Price HKD 2,016,000

2.野生珍珠投资收藏建议
Wild Pearl Investment Collection Suggestions

　　野生珍珠是这几年才在国内流行起来的，其实在欧美国家已经流行了好几个世纪了。在20世纪之前，养殖珍珠还没有发明之前，珠宝首饰上用的都是野生珍珠，所以说野生珍珠一般会出现在古董首饰上。野生珍珠也是一个非常好的投资品，但是投资的主要逻辑是一定要找到一颗品质极佳的野生珍珠。毕竟像那种个头大、皮光好的野生珍珠，真的非常稀有。那些高品质的古董首饰，做工就非常精致，是皇室和贵族所拥有，所以价格肯定不会便宜。所以我个人建议，一定要在自己的预算范围内收入一颗野生珍珠，不着急出售，过几年再看它的价格波动，会是一个非常不错的投资。

　　Wild pearls have only become popular in China in recent years, in fact, they have been popular in Europe and the United States for several centuries. Before the 20th century, before the invention of cultured pearls, wild pearls were used in jewelry, so wild pearls generally appear in antique jewelry. Wild pearls are also a very good investment, but the main logic of investment is to find a wild pearl of excellent quality. After all, wild pearls like that big, shiny leather are really rare. Those high-quality antique jewelry, the workmanship is very exquisite, and is owned by the royal family and nobility, so the price will certainly not be cheap. Therefore, I personally recommend that you must earn a wild pearl within your own budget, and do not rush to sell. Watching its price fluctuations in a few years, it will be a very good investment.

Precious Gemstones as Investment Assets

下篇：名贵珠宝投资篇
Part Two: Luxury Jewelry Investment

第一章
01 Chapter One

名贵珠宝
作为投资资产
Luxury Jewelry
as Investment Asset

一、名贵珠宝投资的历史和发展
The History and Development of Luxury Jewelry Investment

(一) 全球名贵珠宝投资的历史
The History of Global Luxury Jewelry Investment

19世纪末，随着欧洲列强的殖民扩张，许多珍贵的名贵珠宝被带到西方。这导致了名贵珠宝市场的扩大，同时一些富人开始将名贵珠宝作为一种投资手段。

20世纪以来，随着全球贸易的兴起和金融市场的不断发展，名贵珠宝逐渐成为了一个重要的投资市场。人们对名贵珠宝的投资逐渐从传统的装饰性质，发展为更为专业和复杂的资产投资性质。一些大型名贵珠宝公司成立，名贵珠宝拍卖会也开始兴起。名贵珠宝市场交易与拍卖会密切相关。成立于1744年的苏富比是最重要的名贵珠宝拍卖行之一，一般每年举行春秋两季拍卖。委托苏富比拍卖的名贵珠宝包括温莎公爵夫人（Duchess of Windsor）的珠宝珍藏。成立于1766年的佳士得拍卖公司，也是名贵珠宝拍卖的重要拍卖行。1987年，佳士得一颗重64.83克拉的法劳莱钻石拍卖出384万英镑的高价。1986年佳士得在香港的首次拍卖以19世纪和20世纪的绘画及珠宝为主，拍卖总成交额超过1400万港元。此外，名贵珠宝作为另类资产，也是资产配置的重点。在瑞士瑞联、瑞士信贷等国际著名投资机构的家族办公室业务中，除了配置股票、基金等金融产品之外，还积极配置名贵珠宝、艺术品等非金融资产。

By the end of the 19th century, European colonial expansion brought a significant influx of luxury jewelry to the West, leading to an expanded market for luxury jewelry and sparking interest in using it as an investment among wealthy individuals.

Since the 20th century, with the growth of global trade and financial markets, luxury jewelry had evolved into a prominent investment market. Investment in luxury jewelry had transitioned from its traditional decorative nature to a more professional and complex asset class. This shift had led to the establishment of major luxury jewelry companies and a rise in luxury jewelry auctions. Founded in 1744, Sotheby's was one of the most important auction houses for luxury jewelry and generally held sales in the spring and autumn each year. Among the luxury jewels entrusted to Sotheby's was the collection of the Duchess of Windsor. Founded in 1766, Christie's Auction House was also a major auction house in the industry. In 1987, Christie's auctioned a 64.83 carat Falaure diamond for £ 3.84 million. In 1986, Christie's first auction in Hong Kong, focused on 19th - and 20th-century paintings and jewelry, raised more than HK $14 million. Furthermore, luxury jewelry as alternative assets were also playing a significant role in asset allocation. In the family business of well-known international investment institutions such as Swiss Union and Credit Suisse, in addition to allocating financial products such as stocks and funds, they also actively deployed non-financial assets such as luxury jewelry and artworks.

尖晶石胸针

Luxury Jewelry Investment and Appreciation

For example, diamonds are widely used in weddings and religious ceremonies in India; while in the European market, people prefer to buy colored precious stones such as blue precious stones and red precious stones. These market differences provide investors with more choices and opportunities.

进入 21 世纪，名贵珠宝投资市场变得更加全球化和复杂化。名贵珠宝的金融属性进一步凸显，投资者们将名贵珠宝视为一种高价值的投资工具，并购买和交易各种名贵珠宝。除了传统的钻石、红宝石、蓝宝石等名贵珠宝外，人们对于彩色珠宝、珍稀珠宝的需求也逐渐增加。一些投资基金等专业机构也涉足名贵珠宝市场，推动了市场的专业化和规范化。同时，随着科技的发展，人们开始使用各种先进的工具和技术，来评估和鉴定名贵珠宝的质量和价值。例如，光谱分析、硬度测试等方法，被广泛应用于名贵珠宝鉴定，确保投资者们购买到真正的名贵珠宝。此外，全球各地的名贵珠宝市场也呈现出不同的特色。例如，钻石在印度被广泛用于婚礼和宗教仪式；而在欧洲市场，人们更喜欢购买蓝宝石和红宝石等彩色珠宝。这些市场差异，为投资者提供了更多的选择和机会。

小雨老师在中国黄金发表的文章数篇

Entering the 21st century, the luxury jewelry investment market had become more globalize and complex. The financial properties of luxury jewelry were further highlighted, and investors regard luxury jewelry as a high-value investment tool, buying and trading a variety of luxury jewelry. In addition to the traditional diamonds, rubies, sapphires and other luxury jewelry, people's demand for colored jewelry, rare jewelry was also gradually increasing. Some professional institutions such as investment funds were also involved in the luxury jewelry market, which has promoted the professional and standardized development of the market. At the same time, with the development of science and technology, people began to use a variety of advanced tools and techniques to evaluate and authenticate the quality and value of luxury jewelry. For example, spectral analysis, hardness testing and other methods were widely used in the identification of luxury jewelry to ensure that investors buy real luxury jewelry. In addition, the luxury jewelry market around the world also presented different characteristics. For example, diamonds were widely used in weddings and religious ceremonies in India; In the European market, people preferred to buy colored jewelry like sapphires and rubies. These market differences provided investors with more choices and opportunities.

名贵珠宝在欧美国家的财富管理中，已经是成熟的资产配置体系之一。在资产类别中，名贵珠宝投资属于另类资产投资。近年来，以名贵珠宝为代表的另类资产投资火热起来，投资者对于名贵珠宝投资比较理性，既看中名贵珠宝的投资功能，又看中名贵珠宝投资的安全属性，确保全球经济不稳定背景下的资产保值增值。

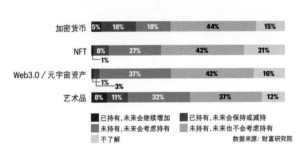

另类投资资产持有情况
资料来源：财富研究院

In the wealth management of European and American countries, luxury jewelry had become one of the mature asset allocation systems. In the asset classification, luxury jewelry investment was an alternative asset investment. In recent years, the investment of alternative assets represented by luxury jewelry had become hot. Investors were more rational in investing in luxury jewelry, focusing not only on the investment function, but also the safety attribute of luxury jewelry investment, to ensure the preservation and appreciation of assets under the background of global economic instability.

（二）中国名贵珠宝投资的历史
The History of China's Luxury Jewelry Investment

改革开放以来，中国经济快速发展。随着人们生活水平的提高，中国人对奢侈品和名贵珠宝的需求也相应增加。这一时期也见证了中国名贵珠宝市场的复苏。名贵珠宝拍卖市场也在中国逐步兴起、发展，大力拓展了名贵珠宝交易市场。

The luxury jewelry auction market had gradually risen and developed in China and had vigorously expanded the luxury jewelry trading market. In June 1995, China Auction Industry Association was established.

1995 年 6 月，中国拍卖行业协会成立。1997 年 1 月，《拍卖法》实施。1993 年 5 月，嘉德公司成立，并开创了中国拍卖行首次拍卖珠宝的记录。2005 年 7 月，保利拍卖公司成立，并迅速拓展了名贵珠宝拍卖业务。然而，和全球名贵珠宝市场相比，中国的名贵珠宝市场及其拍卖市场仍然处于起步壮大阶段。

In January 1997, the Auction Law came into effect. In May 1993, Guardian Company was established and set the record for the first auction of jewelry at a Chinese auction house. In July 2005, Poly Auction Company was established, and quickly expanded the luxury jewelry auction business. However, compared with the global luxury jewelry market, China's luxury jewelry market and its auction market were still in their infancy.

进入 21 世纪，随着中国经济的崛起，越来越多的居民收入增加，提高了人们对名贵珠宝的消费水平，中高收入阶级成为中国名贵珠宝消费主力军。中国成为世界上最大的名贵珠宝市场之一，尤其在钻石、翡翠、红蓝宝石等方面有着很强的需求。随着投资资产配置理念的多元化，结合欧美富人相对比较成熟的资产配置市场（在华尔街的财富管理项目中，富人就是将 5% 的资产作为名贵珠宝投资的），从 2010 年开始，在中国北京、上海、深圳、广州等一线城市，一部分富人开始配置名贵珠宝作为固定资产。香港、深圳、上海，也慢慢有了专业的名贵珠宝交易市场、国际名贵珠宝展。一些名贵珠宝收藏者和投资者，通过国内外的专业名贵珠宝鉴定师、名贵珠宝买手可以获得一些比较名贵的珠宝。

红宝石戒指

In the 21st century, with the rise of China's economy, an growing number of residents' income had increased, which has improved people's consumption level in luxury jewelry. And the middle- and high-income class had become the main force of luxury jewelry consumption in China. China had become one of the world's largest markets for luxury jewelry, especially in diamonds, jadeite, red sapphire and others, with strong demand. With the diversification of investment asset allocation concepts, combined with the relatively mature asset allocation market of rich people in Europe and the United States (in the wealth management project of Wall Street, rich people invest 5% of their assets in luxury jewelry), since 2010, some rich people had begun to allocate luxury jewelry as fixed assets in first-tier cities such as Beijing, Shanghai, Shenzhen and Guangzhou. Hong Kong, Shenzhen, Shanghai, also slowly established professional luxury jewelry trading market and international luxury jewelry exhibition. Some luxury jewelry collectors and investors could obtain some relatively luxury jewelry through professional luxury jewelry appraisers and luxury jewelry buyers at home and abroad.

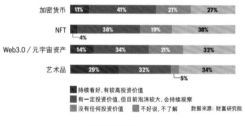

另类投资资产的态度
资料来源：财富研究院

二、名贵珠宝在资产配置中的时机和作用
The Timing and Role of Luxury Jewelry in Asset Allocation

选择名贵珠宝作为资产配置的重要方式，首先要回答三个问题。一是投资名贵珠宝的资产准入是多少？一是名贵珠宝投资在资产配置中的地位如何、占比多少？一是名贵珠宝投资赚钱的可能性和盈利是多少？

To choose luxury jewelry as an important way to allocate assets, you must first answer three questions. Firstly, what is the asset access for investing in luxury jewelry? Secondly, what is the status and proportion of luxury jewelry investment in asset allocation? Last but not least, how much is the possibility and profit of luxury jewelry investment to make money?

(一) 名贵珠宝投资的准入资产是多少?
What are the Entry Assets for Luxury Jewelry Investment?

这个问题的核心,是投资者在什么时候才能投资名贵珠宝。也就是说,投资者在什么时候才能拥有足够的资产,去配置名贵珠宝。我给大家列个很简单的标准,当你每年至少有 50 到 100 万的稳定闲钱,就可以慢慢做名贵珠宝的投资,名贵珠宝是一个小而精的资产,是需要用时间去换价值的产物,你要做好至少 3 年、5 年、10 年,甚至一代人的长期规划。它不是核心房地产,需要一开始投大几百万元的大资金;它也不是黄金,500 元就可以买一点点。为什么我说需要至少 50 万元,因为只有这样的价格,你才能买到一颗有保值能力的名贵珠宝。每年你可以买进卖出,你的名贵珠宝会交易得越来越大,越来越名贵保值,甚至可以进佳士得、苏富比拍卖。如果你的闲置金钱达到上千万元上亿元,你的配置就可以更多元化。

The key to the question is that, when investors can invest in luxury jewelry? In other words, when will investors have enough assets to allocate on luxury jewelry. Let me list a very simple standard for you. When you have at least 0.5 to 1 million stable spare money every year, you can slowly start to invest in luxury jewelry. Luxury jewelry is a small and fine asset, it is the kind of product that takes time to exchange for value. You will have to make at least 3 years, 5 years, 10 years, or even a generation's long-term plan. It is not a core industry like real estate that requires a large investment of several million yuan at the beginning. And it's not gold, you can buy a little bit for 500 yuan. Why I say you will need at least 500,000 yuan? That is because only at this price you can buy a luxury jewelry that has the ability to preserve its value. Each year you can buy and sell, and your luxury jewelry will trade bigger and bigger, keeping its value, and even be auctioned at Christie's and Sotheby's. If your idle money reaches tens of millions dollars, your allocation can be more diversified.

对于高收入家庭,可以更多考虑配置名贵珠宝的投资目的。例如,可以在青年时期配置名贵珠宝用于婚恋,在中年阶段配置名贵珠宝提升资产收益,在老年时期配置名贵珠宝用于财产保值增值、财富传承和情感传递。

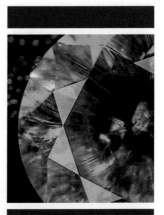

翠榴石

For high-income families, more consideration can be given to the investment purpose of allocating luxury jewelry. For example, you can allocate luxury jewelry in youth for marriage, allocate luxury jewelry in middle age to enhance asset returns, and allocate luxury jewelry in old age for property preservation and appreciation, wealth inheritance and emotional transmission.

名贵珠宝的投资门槛应不同需求而异。相对于其他投资方式,如房地产、艺术品等,名贵珠宝的投资门槛相对较低,价格具有浮动性。投资者可以根据自己的经济实力选择不同的投资级别和规模。可以从小规模的名贵珠宝投资开始,逐步积累财富。名贵珠宝投资具有稳健性和多样性,能够满足不同投资者的需求。

The investment threshold of luxury jewelry should vary according to different needs. Compared with other investment methods, such as real estate, art pieces, etc., the investment threshold of luxury jewelry is relatively low, and the price is floating. Investors can choose different investment grades and sizes according to their economic strength. You can start with a small investment in luxury jewelry and build up your wealth. Luxury jewelry investment is robust and diverse and can meet the needs of different investors.

(二) 名贵珠宝投资在资产配置中的地位如何?
What is the Status of Luxury Jewelry Investment in Asset Allocation?

名贵珠宝投资在资产配置中属于另类投资。虽然名贵珠宝在富裕家庭和高净值家庭的资产配置中占比不高,但是,名贵珠宝的保值增值空间大。名贵珠宝投资主要关系到居民资产配置的阶段性财务目标,这需要充分认识资产配置的三种主要功能,包括保险箱安全功能、税后收入增值功能、钱生钱补缺功能。

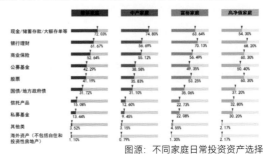

图源:不同家庭日常投资资产选择
资料来源:2022~2023中国家庭资产配置白皮书
White Paper on Household Asset Allocation in China from 2022 to 2023

Luxury jewelry investment is an alternative investment in asset allocation. Although the proportion of luxury jewelry in the asset allocation of wealthy families and high net worth families is not high, but the value of luxury jewelry is large. Luxury jewelry investment is mainly related to the phased financial goals of residents' asset allocation, which requires a full understanding of the three main functions of asset allocation, including safe-deposit box security function, value-added function of after-tax income, and money-generating function.

资产配置的保险箱安全功能，主要是以资产保障家庭生活质量。这要求第一，资产配置不追求高收益。第二，要求具有流动性，资产配置能够保障家庭半年左右的开支，用于支付房贷、车贷、保险费、信用卡和日常生活开支等等。第三，要求具有兜底能力，在进行资产配置时，保险必不可少。

The safe-deposit box security function of asset allocation mainly guarantees the quality of family life with assets. This requires, first, that asset allocation does not seek for high returns. Second, it is required to have liquidity, and asset allocation can guarantee the expenditure of the family for about half a year, which is used to pay for mortgages, car loans, insurance, credit cards and daily living expenses. Third, it requires the ability for overall protection, and insurance is essential when carrying out asset allocation.

资产配置的税后收入增值功能，这部分收入构成中产阶层居民的核心收入来源。第一，这要求居民强制储蓄，每月固定储蓄一定金额，从而保障家庭日常支出。第二，要求定期投资，居民将多余财产用于投资，以确保资产增值。第三，要求一定的收益率，居民通过不断学习理财知识，提升自我理财能力，获得更多赚钱的能力。

The value-added function of after-tax income of assets allocation, which constitutes the core income source of middle-class residents. First, it requires residents to save compulsively, saving a fixed amount each month, so as to protect the household's daily expenses. Second, regular investment is required, and residents invest their excess property to ensure asset appreciation. Third, a certain rate of return is required. Residents can improve their self-financing ability by constantly learning financial knowledge and gain more earning ability.

小雨老师在美国GIA宝石实验室考察

资产配置的钱生钱补缺功能，主要是以资产收益来为保险箱安全功能补缺。第一，这要求资产配置获得长期稳定的、较高的收益率。第二，要求具有一定的流动性，通过中长期投资降低资产收益波动性，从而实现家庭资产配置增值的阶段性财务目标。第三，要求具有安全性，资产配置要设置止损线，在止损线以内，通过牺牲一定的安全性来获得较高收益率。

The ability of generating money of assets allocation to make up for the deficiency function, which mainly uses the asset income to make up for the safety function of the safe deposit box. First, it requires asset allocation to achieve long-term stability and higher returns. Second, it is required to have a certain liquidity, and reduce the volatility of asset income through medium and long-term investment, so as to achieve the phased financial goal of household asset allocation appreciation. Third, the requirement of security, asset allocation to set a stop loss line. Within the stop loss line, by sacrificing a certain level of security, a higher rate of return could be obtained.

从资产配置的主要功能看，名贵珠宝属于其他类的另类资产配置，主要是发挥税后收入增值功能和钱生钱补缺功能。也就是说，在经济稳健增长时期，名贵珠宝投资能够获得较高的增值空间。在经济低迷时期，名贵珠宝投资能够帮助投资者在不确定性中找到确定性，获得较好的稳定收益，提升家庭资产配置的财务水平。从中长期投资来看，名贵珠宝具有良好的价格上涨空间，可以充分保障资产的安全性和收益性。

Seeing from the aspects of the main functions of asset allocation, luxury jewelry belongs to other types of alternative asset allocation, which mainly plays the value-added function of after-tax income and the function of generating money to make up for the deficiency. In other words, in the period of steady economic growth, luxury jewelry investment can obtain a higher value-added space. In the economic downturn, luxury jewelry investment can help investors find certainty in the uncertainty, obtain a better stable income, and improve the financial level of household asset allocation. From the perspective of medium and long-term investment, luxury jewelry has good room for price rise, which can fully protect the safety and profitability of assets.

（三）名贵珠宝投资赚钱的可能性和盈利有多少？
How much is the Possibility and Profit of Luxury Jewelry Investment to Make Money?

这里涉及投资的胜率和赔率。所谓胜率，就是投资者在每次交易时能够盈利的概率，通俗点说就是赚钱的把握，重点在于盈利的确定性。所谓赔率，就是投资者每次平均的投资盈利和投资亏损之间的比值大小，通俗点说就是赚了多少钱，重点在于投资的盈利水平。我们要从比较的角度，来看名贵珠宝投资的胜率与赔率。

This involves the winning rate and odds of investing. The so-called winning rate refers to the probability that investors can make a profit in each transaction. In layman's terms, it is the certainty of making money. The focus is on the certainty of profit. The so-called odds refer to the ratio between an investor's average investment profit and investment loss each time. In layman's terms, it means how much money is earned. The focus is on the profitability of the investment. We want to look at the winning rate and odds of precious gemstone investment from a comparative perspective.

小雨老师讲座

第一，低胜率、低赔率的投资。投资者选择业余炒股、创业等投资，这属于低胜率、低赔率的投资方式。这种投资方式，获得交易盈利的可能性较小，收益水平较低。不建议投资者选择这种投资方式，当然可以轻仓投资。

第二，低胜率、高赔率的投资。投资者选择彩票、天使投资、保险等投资，这属于低胜率、高赔率的投资方式。这种投资方式，获得交易盈利的可能性较小，但是收益水平相对较高。不建议投资者重仓这种投资，可以轻仓投资或者直接回避。

第三，高胜率、低赔率的投资。投资者选择银行存款、基金定投、固定收益、打工等投资，这属于高胜率、低赔率的投资方式。这种投资方式，获得交易盈利的可能性较高，但是收益水平相对较低。保守型的投资者可以选择这种投资方式，以获得较为确定性的收益。

第四，高胜率、高赔率的投资。投资者在2000年到2019年选择投资茅台股票，或者投资房地产，这属于高胜率、高赔率的投资方式。这种投资方式，获得交易盈利的可能性较高，收益水平也相对较高。这是过去20年投资者重仓的投资方式。

Firstly, low win rate, low odds investment. Investors choose amateur stocks, entrepreneurship and other investment, which belongs to the low win rate, low odds investment. This way of investment, the possibility of obtaining trading profits is small, and the level of income is low. It is not recommended for investors to choose this investment method, of course, you can invest lightly.

Secondly, low win rate, high odds investment. Investors choose lottery, angel investment, insurance and other investments, which belong to the low win rate, high odds investment method. This way of investment, the possibility of obtaining trading profits is small, but the level of return is relatively high. It is not recommended for investors to invest heavily in this kind of investment, which can be lightly invested or directly avoided.

Thirdly, investment with high winning rate and low odds. Investors choose bank deposits, fund investment, fixed income, work and other investments, which belong to the investment method of high win rate and low odds. This investment method has a high possibility of obtaining trading profits, but the level of return is relatively low. Conservative investors can choose this type of investment to obtain a more certain income.

Fourthly, investment with high winning rate and high odds. Investors choose to invest in Moutai stocks from 2000 to 2019, or invest in real estate, which belongs to this investment method with high win rate and high odds. This way of investment, the possibility of obtaining trading profits is high, and the level of income is relatively high. This is the way investors have invested heavily over the past 20 years.

但是，当经济低迷、潮水退去，选择何种资产才能保证较高的盈利可能性，获得较高的收益水平，这非常考验投资者的远见和判断能力。在投资渠道较少的时期，名贵珠宝投资是高胜率、高赔率的有效投资方式。从投资胜率来看，名贵珠宝每次投资赚钱的把握比较大，再次交易盈利的概率比较高，投资名贵珠宝获得较高收益具有确定性。从赔率来看，名贵珠宝投资的赚钱能力也相对较高，赚多赔少，能够获得较高的收益水平。

However, when the economy is in a downturn and the tide is receding, which assets can be selected to ensure a higher profit possibility and obtain a higher income level. This is a harsh test of investors' vision and judgment. In the period of less investment channels, luxury jewelry investment is an effective way of investment with high win rate and high odds. From the perspective of investment success rate, the grasp of making money in each investment of luxury jewelry is relatively large, and the probability of making profits in trading again is relatively high, and it is certain that investment in luxury jewelry can obtain higher returns. From the point of view of the odds, the earning power of luxury jewelry investment is also relatively high, earn more and lose less, and can obtain a higher level of income.

三、名贵珠宝投资市场及其吸引力
Luxury Jewelry Investment Market and Its Attractiveness

资产配置的主要作用是规避经济金融风险，让投资者的资产保值增值。如何选择资产配置方式非常重要。一方面，要防止"黑天鹅事件"这种小概率、难预测的突发风险导致的资产损失。另一方面，防止"灰犀牛事件"这种大概率、可预测、波及范围大的风险导致的资产损失。选择名贵珠宝投资，能够通过积极的资产配置筹划，帮助投资者提升资产的收益率。投资名贵珠宝，可帮助投资者抵御通货膨胀、货币贬值风险，抵御经济周期风险，助力超高净值家族财富管理。

The main function of asset allocation is to avoid economic and financial risks, so that investors can maintain and increase their assets. How you choose your asset allocation is very important. On the one hand, it is necessary to prevent asset losses caused by "black swan events" such small probability and unpredictable sudden risks. On the other hand, to prevent the "gray rhino event" such large probability, predictable, large-scale risk causing the loss of assets. Choosing luxury jewelry investment can help investors to improve the return on assets through active asset allocation planning. Investing in luxury jewelry can help investors resist the risk of inflation, currency depreciation, withstand the risk of the economic cycle, and help on ultra-high net-value family wealth management.

（一）名贵珠宝投资需求上升
Investment Demand for Luxury Jewelry Is Rising

名贵珠宝投资需求逐步上升，名贵珠宝的投资优势更为明显。总体而言，中国 GDP 仍保持较强的增速，这确保了名贵珠宝消费和投资的整体增长趋势。中国人均 GDP 和人均可支配收入的增长，提升了名贵珠宝投资消费的能力。人均可支配收入增长和名贵珠宝投资消费能力提升之间，往往呈现出正相关的增长关系，增长轨迹基本相同。

The investment demand for luxury jewelry is gradually rising, and the investment advantage of luxury jewelry is more obvious. Overall, China's GDP still maintains a strong growth rate, which ensures the overall growth trend of luxury jewelry consumption and investment. The growth of China's per capita GDP and per capita disposable income has enhanced the ability to invest and consume luxury jewelry. There is often a positive correlation between the growth of per capita disposable income and the improvement of investment and consumption power of luxury jewelry, and the growth trajectory is basically the same.

在 2011 年以前，美国是全球最大的名贵珠宝投资消费市场，但是自 2011 年以来，中国名贵珠宝消费市场占据全球第一的位置。值得注意的是，珠宝是一个大类，名贵珠宝是这个大类里的一小部分，名贵的才是值得投资的。虽然国内名贵珠宝投资市场在一线市场的富人阶层逐渐启蒙，但它仍然相对较小，相比黄金、股票等传统投资工具而言，名贵珠宝投资的知名度和市场规模仍有很大的上升空间。

Before 2011, the United States was the world's largest luxury jewelry investment consumer market, but since 2011, China's luxury jewelry consumer market had occupied the first position in the world. It is worth noting that jewelry is a big category, luxury jewelry is a small part of this big category, luxury is what worth investing in. Although the domestic luxury jewelry investment market is gradually enlightened by the wealthy class in the first-tier market, it is still relatively small. Compared with traditional investment tools such as gold and stocks, the popularity and market size of luxury jewelry investment still has a lot of room to rise.

（二）名贵珠宝投资提供了更多选择性
Luxury Jewelry Investment Provides More Options

居民的家庭资产通常包括现金、存款、股票、基金、理财、期货等金融资产，也包括住宅、商铺、厂房、名贵珠宝、古董、艺术品等非金融资产。名贵珠宝投资给居民的资产配置提供了多样选择可能性。名贵珠宝除了自身的稀有属性、审美属性、情感寄托属性之外，更重要的是具备金融属性。名贵珠宝是资产投资的重要载体，能够将货币价值转化为稀有的名贵珠宝商品价值，并在未来获得较高的投资收益。1997年1月《拍卖法》实施。1993年5月嘉德公司成立，并开创了中国拍卖行首次拍卖珠宝的记录。2005年7月保利拍卖公司成立，并迅速拓展了名贵珠宝拍卖业务。然而，和全球名贵珠宝市场相比，中国名贵珠宝市场及其拍卖市场仍然处于起步壮大阶段。

Household assets usually include financial assets such as cash, deposits, stocks, funds, wealth management and future goods, as well as non-financial assets such as housing, shops, factories, luxury jewelry, antiques and works of art. Luxury jewelry investment provides a variety of options for residents' asset allocation. In addition to their own rare attributes, aesthetic attributes, emotional sustenance attributes, the most important is to have financial attributes. Luxury jewelry is an important carrier of asset investment, which can transform the monetary value into the value of rare luxury jewelry commodities and obtain higher investment returns in the future. In January 1997, the Auction Law came into effect. In May 1993, Guardian Company was established and set the record for the first auction of jewelry at a Chinese auction house. In July 2005, Poly Auction Company was established and rapidly expanded its luxury jewelry auction business. However, compared with the global luxury jewelry market, China's luxury jewelry market and its auction market are still in the initial stage of growth.

名贵珠宝投资符合未来资产配置的新趋势，按照LDEP投资法则投资名贵珠宝，更能让资产保值增值。按照日本金融机构在低利率时代的投资策略，名贵珠宝投资的L，即long，意思是拉长周期，名贵珠宝可以延长投资期限来获得未来更高的投资收益。名贵珠宝投资的D，即diversified，意思是多元化，名贵珠宝在另类资产投资中属于优质资产，可以通过投资多元化来降低资产风险。名贵珠宝投资的E，即external，意思是投资海外，名贵珠宝是全球硬通货，可以作为海外资产来分散投资风险。名贵珠宝的P，即passive，意思是被动化，名贵珠宝可以实现"睡后增值"，通过被动投资来获得更加稳定的资产投资收益。

绝地尖晶石戒指

Luxury jewelry investment is in line with the new trend of future asset allocation, and investing in luxury jewelry according to the LDEP investment rule can better preserve and increase the value of assets. According to the investment strategy of Japanese financial institutions in the era of low interest rates, the L, that is, long, means that the investment period of luxury jewelry can be extended to obtain higher investment returns in the future. The D of luxury jewelry investment is diversified, which means diversification. Luxury jewelry belongs to high-quality assets in alternative asset investment, and asset risks can be reduced through investment diversification. The E of luxury jewelry investment, referring to external, means to invest overseas. Luxury jewelry is a global hard currency, can be used as overseas assets to diversify investment risks. P of luxury jewelry is passive, which means passivity. Luxury jewelry can realize "value increase after sleep" and obtain more stable asset investment income through passive investment.

名贵珠宝消费在居民消费支出中，属于可选择性消费支出，具有很强的弹性空间。这表明，居民的名贵珠宝消费和投资意愿，经常受到宏观经济发展和居民收入水平的影响。在经济发展持续稳健增长的时期，以及在居民消费等级提升的条件下，奢侈品市场前景更为可观，居民选择名贵珠宝消费和投资的需求会增长。但是，在经济较为低迷的时期，对于高净值收入的家庭而言，名贵珠宝依旧是保值增值的重要投资方式。

Among the residents' consumption expenditure, the consumption of luxury jewelry belongs to the optional consumption expenditure, which has a strong flexible space. This shows that residents' willingness to consume and invest in luxury jewelry is often affected by macroeconomic development and residents' income level. In the period of sustained and steady economic development, as well as the condition of residents' consumption level improvement, the luxury market prospects are more promising, and the demand of residents to choose luxury jewelry consumption and investment will increase. However, in the period of economic downturn, for high net-value families, luxury jewelry is still an important investment way to preserve and increase value.

（三）名贵珠宝投资抵御货币贬值风险
Luxury Jewelry Investment to Resist the Risk of Currency Depreciation

选择名贵珠宝投资，是抵御宏观经济中通货膨胀、货币贬值的必要手段。宏观经济和所有资产价格紧密相关，是一切资产定价的前提，宏观经济决定了所有资产的涨跌。宏观经济关注的重点，是经济增长和通货膨胀。中国经济要走出疫情困境和海外加息影响，而国外的难点在于控制较高的通货膨胀。当前，受到全球货币宽松、货币贬值的影响，全球内生通胀压力逐步形成。展望未来，全球经济发展前景依然具有下行风险，通货膨胀压力会比较持久。受粮食危机、能源不确定性和地缘政治风险影响，全球通胀率难以下降。从长远角度看，适度的通货膨胀也是促进经济增长的重要方式，但是通货膨胀也将导致居民财富缩水。

Choosing luxury jewelry investment is a necessary means to resist inflation and currency depreciation in the macro economy. The macro economy is closely related to all asset prices and is the premise of all asset pricing. The macro economy determines the rise and fall of all assets. The focus of Macroeconomic is on economic growth and inflation. China's economy has to get out of the dilemma of the epidemic and the impact of overseas interest rate hikes, while the difficulty abroad is to control high inflation. At present, under the influence of global monetary easing and currency depreciation, global internal inflation pressure has gradually formed. Looking ahead, the future foresee for global economic development still has downside risks, and inflationary pressure will be relatively persistent. The food crisis, energy uncertainty and geopolitical risks are making it difficult for global inflation to fall. In the long run, moderate inflation is also an important way to promote economic growth, but inflation will also lead to a decline in people's wealth.

蓝宝石戒指

（四）名贵珠宝提升通缩时期投资收益率
Luxury Jewelry to Improve Investment Returns in the Period of Deflation

当前，国内经济通缩压力较大，导致经济负向循环风险上升，投资渠道减少。具体而言，物价水平低迷导致企业利润空间缩小，企业盈利困境削减投资和就业，就业压力加大以及失业率上升导致居民收入增速降低，居民收入降低导致消费需求不足，进一步导致物价水平偏低。经济通缩的最终结果，是资产投资渠道减少，资产保值压力较大，甚至出现资产缩水。

At present, the deflationary pressure of the domestic economy is relatively large, leading to the rise of the risk of negative economic circulation and the reduction of investment channels. Specifically, the low price level leads to the reduction of enterprise profit space. The difficulty of enterprise profit reduces investment and employment, the increase of employment pressure and the rise of unemployment rate lead to the decrease of residents' income growth, the decrease of residents' income leads to insufficient consumer demand, and further leads to the low price level. The ultimate result of economic deflation is the reduction of asset investment channels, greater pressure on asset preservation, and even asset shrinkage.

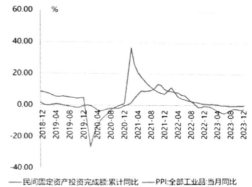

PPI持续为负，民间投资增速下行
资料来源：中诚信国际研究院，Wind
PPI Remained Negative and Private Investment Growth Declined
Source: China Integrity International Research Institute, Wind

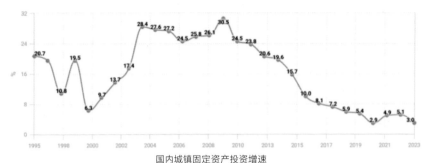

国内城镇固定资产投资增速
资料来源：CEIC、国泰君安证券
Growth Rate of Domestic Urban Fixed Asset Anvestment
Source: CEIC, Guotai Junan Securities

在投资渠道明显减少的经济通缩时期，稳健的投资渠道对于保障资产收益率非常重要。名贵珠宝在另类资产投资中具有明显优势，投资渠道稳定，资源稀缺，价格上升空间强劲，在全球买卖交易便利，能够满足投资者在经济通缩期间的稳健投资需求，使资产投资不缩水、保值增值。

In the period of economic deflation when investment channels are significantly reduced, a sound investment channel is very important to ensure the return on assets. Luxury jewelry has obvious advantages in alternative asset investment, stable investment channels, scarce resources, strong room for price rise, and convenient global trading, which can meet investors' steady investment demand during economic deflation, so that asset investment will not shrink, keep flat and even increase value.

（五）名贵珠宝重塑存量时代的投资结构
Luxury Jewelry Reshapes the Investment Structure of the Stock Era

投资属于认知的变现，资产投资需要把握时代性，找到顺应时代发展趋势的投资。当前，经济正在发生时代性的变化，经济发展已经从快速发展的增量时代，进入了减速发展的存量时代。从打破旧秩序到重建新秩序的时期，如何选择正确的资产投资渠道是非常困难的事情，这非常考验投资者的认知。

Investment belongs to the realization of cognition, and asset investment needs to grasp the times and find the investment that conforms to the development trend of The Time. At present, the economy is undergoing epochal changes, and the economic development has entered the stock era of decelerating development from the era of rapid growth and increment. From the period of fighting the old order to rebuilding the new order, how to choose the right asset investment channel is a very difficult thing, which is a harsh test on the cognition of investors.

每个时代都有自己的核心资产。在经济增量时代，资产配置面对的是系统性机会，经济总量特征显著，炒房地产比较赚钱，股票、信托、理财、地产基建等产业是核心资产，投资估值并不需要考虑资产安全边际，经济增长可以弥合投资的局部亏损，投资者可以靠胆量制胜。但是，进入经济存量时代，资产配置面对的是系统性风险，资产结构重于一切，减速换挡去杠杆成为主旋律，金融杠杆会使资产随时变为负债，居民消费能力、消费结构和消费愿望都发生深层次变革，风险控制成为首要任务，投资者需要靠专业制胜。

Each era has its own core assets. In the era of economic increment, asset allocation was faced with systemic opportunities, the characteristics of economic aggregate were significant, real estate speculation was more profitable, and stocks, trust, financial management, real estate infrastructure and other industries were core assets. Investment valuation did not need to consider the safety margin of assets, economic growth could bridge the partial loss of investment, and investors could rely on courage to win. However, entering the era of economic stock, asset allocation is facing systemic risk, asset structure is more important than everything. Slowing down, shifting gears, and deleveraging have become the main theme. Financial leverage will make assets into liabilities at any time, residents' consumption capacity, consumption structure and consumption desire have undergone profound changes. Risk control has become the top priority, and investors need to rely on professional victory.

祖母绿戒指

因此，在经济存量时代，必须找到价值投资，把合理的估值作为投资的前提条件。名贵珠宝属于价值投资种类，是对时代发展趋势的投资，可以和长期持股核心资产画等号，类似于锂矿石、稀土等核心资产，一方面极其稀缺而有更高价值，另一方面是在全球可便利交易的一般等价物。从生命周期看，名贵珠宝几乎没有生命周期衰退期，没有类似于其他行业的衰退期估值陷阱，不会使投资者陷入泡沫时期的被动投资导致资产缩水。从专业性看，名贵珠宝投资极具专业性，能够适应存量时代专业制胜的投资要求，能够发掘常人无法认知的价值，掌握更高的资产定价权。名贵珠宝投资具有低风险属性、高收益率特征，周期性波动较小、能够跑赢通胀，属于成长型核心资产，能够满足居民资产配置的低风险和较高收益需求。

Therefore, in the era of economic stock, we must find value investment and take reasonable valuation as a prerequisite for investment. Luxury jewelry belongs to the category of value investment, which is an investment in the development trend of The Times, and can be equated with long-term holding core assets, similar to core assets such as lithium ores and rare earths. On the one hand, they are extremely scarce and have higher value, and also general equivalents that can be easily traded around the world on the other hand. From the life cycle point of view, luxury jewelry has almost no life cycle recession period. There is no recession valuation trap similar to other industries, and investors will not fall into the passive investment during the bubble period resulting in asset shrinkage. From the professional point of view, luxury jewelry investment is very professional, which can adapt to the stock era of professional winning investment requirements, and can explore the value that ordinary people not recognize, and can master higher asset pricing power. Luxury jewelry investment has the characteristics of low risk, high yield, small cyclical fluctuations, outperforming inflation, and is a growth core asset. It can meet the needs of low risk and high return of residents' asset allocation.

（六）名贵珠宝投资抵御经济周期风险
Luxury Jewelry Investment to Resist Economic Cycle Risks

小雨老师跟韩旭女士交流宝石

为什么在投资中要配置名贵珠宝，这需要从大历史观的角度，从总体上把握经济周期与资产配置之间的关系。

Why to allocate luxury jewelry in investment? It needs to reviewed by the relationship between the economic cycle and asset allocation in general from the perspective of a large historical view.

从历史上看，经济发展往往要经历经济衰退时期、经济复苏时期、经济过热时期和经济滞涨时期，而不同时期直接影响了资产配置的策略。在经济复苏时期和经济过热时期，包括名贵珠宝在内的大类资产配置，一般都能够获得较好的资产收益。但是，在经济衰退时期和经济滞涨时期，经济增长速度下降，甚至出现经济通缩的情况，经济发展从"防过热"转向"防过冷"，导致资产投资渠道明显减少，资产不但不能保值，甚至有缩水的风险。

Historically, economic development often goes through periods of economic recession, economic recovery, economic overheating and economic stagflation. The different periods directly affect the strategy of asset allocation. During the period of economic recovery and economic overheating, the asset allocation of major categories, including luxury jewelry, is generally able to obtain better asset returns. However, in the period of economic recession and economic stagflation, the economic growth rate declines, and even the situation of economic deflation, economic development from "prevent overheating" to "prevent too cold", resulting in significantly reduced asset investment channels. The assets not only can not maintain value, and even have the risk of shrinking.

因此，如何选择有效的资产配置至关重要。面对经济增长下行的宏观环境，能够穿越经济周期的资产类别将是资产配置的优先选项。投资者应该选择具有长期增值能力的投资方式。名贵珠宝单位价值高、溢价能力强，是资产配置中的有效资产，能够抵御经济下行周期的风险，抵御经济通缩导致资产缩水的风险，提升资产配置的收益率。

Therefore, how to choose effective asset allocation is crucial. In the face of a macro environment of downward economic growth, asset classes that can travel through the economic cycle will be the preferred option for asset allocation. Investors should choose an investment method that has the ability to increase long-term value. Luxury jewelry is an effective asset in asset allocation with high unit value and strong premium ability, which can resist the risk of economic downturn cycle, resist the risk of asset shrinkage caused by economic deflation, and improve the return rate of asset allocation.

四、名贵珠宝投资与高净值家族财富管理
Luxury Jewelry Investment and High Net-value Family Wealth Management

（一）家族财富配置的类型与名贵珠宝投资
The Types of Family Wealth Allocation and Luxury Jewelry Investment

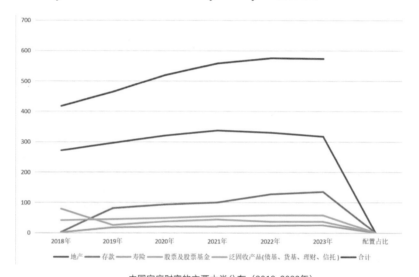

中国家庭财富的主要大类分布（2018-2023年）
单位：万亿元人民币
资料来源：Wind，中金财富
Distribution of Major Categories of Household Wealth in China (2018-2023), Unit: Trillion Yuan
Credit: Wind, CICC Wealth

当前，中国家族财富的投资大类，主要分布在地产、存款、寿险、股票及基金、泛固收产品等领域。

At present, the investment categories of Chinese family wealth are mainly distributed in real estate, deposits, life insurance, stocks and funds, pan-fixed income products and other fields.

中国高净值资产家庭也积极布局海外投资，主要投资地区是欧洲、北美和亚洲其他国家，主要投资币种是美元、欧元，主要投资类型有股票、基金、信托等跨境金融，以及医疗健康、金融、企业服务、区块链、文娱传媒、先进制造、地产等跨境产业。相比之下，海外投资布局艺术品、名贵珠宝等另类资产的相对较少，部分资管机构的家族财富管理业务，会涉及名贵珠宝等另类资产。从趋势看，如何实现资产配置的保值增值，才是高净值资产家庭在海外资产布局的重点。

Chinese high net-value households are also actively investing overseas, mainly in Europe, North America and Asia. The main investment currencies are US dollar and Euro. The main investment types include cross-border finance such as stocks, funds and trusts, as well as cross-border industries such as healthcare, finance, corporate services, blockchain, culture, entertainment and media, advanced manufacturing and real estate. In contrast, overseas investment in alternative assets such as artworks and luxury jewelry are relatively small. Partially of the family wealth management business of asset management institutions will involve alternative assets such as luxury jewelry. From the perspective of the trend, how to realize the preservation and appreciation of asset allocation is the focus of high net value families in overseas asset layout.

从资产配置方面看，名贵珠宝将会成为未来家庭资产配置的重要方式。从长期视角看，未来名贵珠宝市场将从之前的供给主导型市场，转变为未来长期维度的需求主导型市场，个性化的高净值名贵珠宝，将逐步成为资产配置的重要选择。

From the perspective of asset allocation, luxury jewelry will become an important way of family asset allocation in the future. From a long-term perspective, the future luxury jewelry market will change from the previous supply-led market to the future long-term dimension of demand-led market, personalized high net value luxury jewelry, will gradually become an important choice for asset allocation.

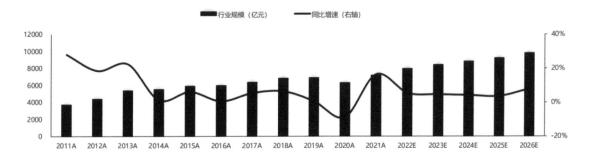

中国内地黄金名贵珠宝行业市场规模预测
资料来源：欧睿数据，民生证券研究院
Market Size Forecast of China Mainland's Gold and Luxury Jewelry Industry
Source: Euromonitor Data, Minsheng Securities Research Institute

（二）以名贵珠宝投资降低家族财富配置的风险
Invest in Luxury Jewelry to Reduce the Risk of Family Wealth Allocation

名贵珠宝投资能够规避资产减损风险。中国富裕家庭及其拥有的财富逐年增长，在私人财富快速增长并代际传承的过程中，做好顶层设计，是超高净值家族财富管理的重要目标。在家族财富传承过程中，要规避类似 P2P 暴雷、股价下跌、上市公司债务问题等风险。名贵珠宝投资具有保值增值的功能，能够助力家族财富有效规避上述风险。

Luxury jewelry investment can avoid the risk of asset impairment. China's wealthy families and their wealth are increasing year by year. In the process of rapid growth of private wealth and intergenerational transmission, a good top-level design is an important goal of wealth management for ultra-high net value families. In the process of family wealth inheritance, it is necessary to avoid risks such as P2P trap, stock price decline, and debt problems of listed companies. Luxury jewelry investment has the function of preserving and increasing value, which can help family wealth effectively avoid the above risks.

在家族财富管理中，名贵珠宝是非常重要的资产配置方式。如何使高净值资产家庭优化投资资产，关键是选准投资工具。名贵珠宝投资，通常要在全世界范围内选择名贵珠宝，这可以帮助家族财富实现全球资源配置。名贵珠宝投资，能够满足超高净值资产家庭的财富传承、财富安全、财富保值、财富增值需求，实现积累的物质财富的代际传承，并以名贵珠宝为媒介实现精神财富的传承。

In family wealth management, luxury jewelry is a very important way of asset allocation. How to optimize the investment assets of high net value households, the key is to select the right investment tools. Luxury jewelry investment, which usually involves selecting luxury jewelry around the world, can help the family wealth to achieve global resource allocation. Luxury jewelry investment can meet the needs of ultra-high net value families for wealth inheritance, wealth security, wealth preservation and wealth appreciation, realize the intergenerational inheritance of accumulated material wealth, and realize the inheritance of spiritual wealth with luxury jewelry as the medium.

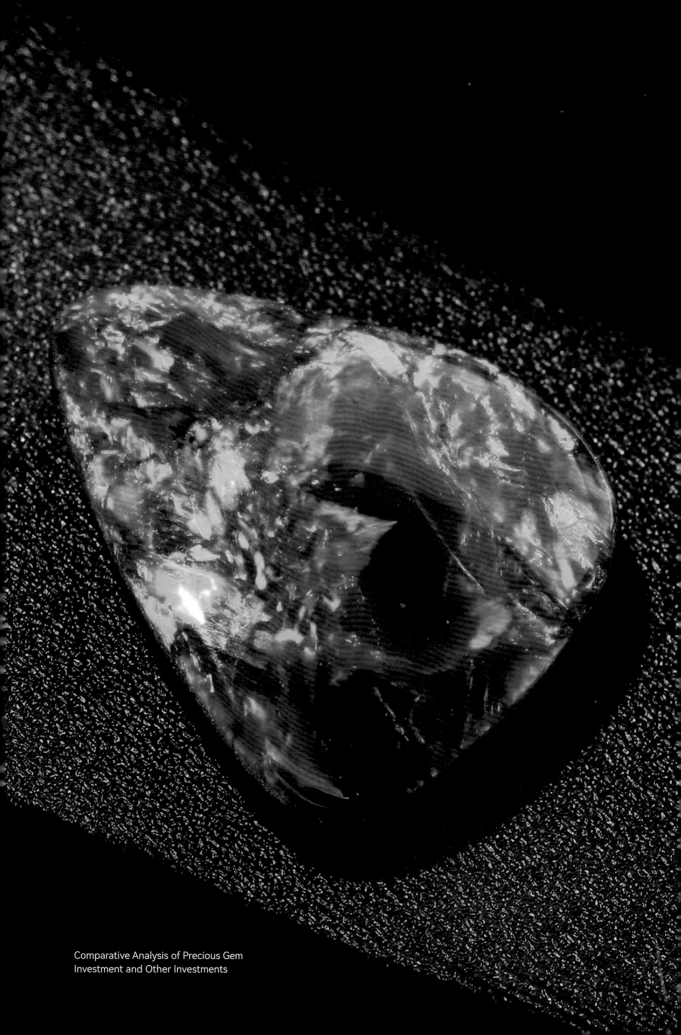

Comparative Analysis of Precious Gem
Investment and Other Investments

第二章
02 Chapter Two
名贵珠宝投资与其他投资的对比分析
Comparative Analysis of Luxury Jewelry Investment and Other Investments

名贵珠宝投资与黄金投资、股市投资、房地产投资、基金投资、理财等，都是金融投资领域的重要组成部分。但是，它们在投资标的、价值评估、交易方式、风险与回报以及专业性等方面，都存在显著差异。

　　首先，谈谈宏观经济对股市、房地产、基金、理财、保险的影响。这五类资产投资与宏观经济的涨跌是顺周期关系，很难穿越经济周期实现资产保值增值。当宏观经济形势较好并持续上涨的时候，这五类资产投资的价格和收益通常会持续上涨，当宏观经济形势较差并持续下滑的时候，这五类资产的价格和收益通常会持续下跌。例如，房地产和股市的涨跌就是反映宏观经济形势涨跌的晴雨表。

　　其次，谈谈名贵珠宝投资、黄金投资与宏观经济的关系。名贵珠宝投资与股市、房地产、基金、理财、保险的投资逻辑不同，名贵珠宝投资能够穿越经济周期，通过逆周期调节实现资产保值增值。在经济通胀的时候，名贵珠宝投资能够对冲通胀风险，稳定资产价格并实现增值。在经济通缩时期，投资渠道减少，名贵珠宝则提供了优质的投资标的，从而确保并提高资产收益率。此外，名贵珠宝投资主要是在高净值客户中流通交易，准入门槛较高，不受大众收入涨跌的影响。相较而言，黄金投资也具有保值增值的功能，但是黄金投资准入门槛较低，受大众收入涨跌的影响较大，跨境流通性较小，在确保和提高资产收益率方面明显弱于名贵珠宝。

　　当前，股票下行、房地产价格下行、基金投资收益下降、理财规模收缩，使居民资产增速显著回落。相较而言，名贵珠宝投资更具保值、便利、低风险、高收益等特征，能够使资产保值增值。

　　Luxury jewelry investment, together with gold investment, stock market investment, real estate investment, fund investment, wealth management, etc., are important parts of the financial investment field. However, there are significant differences between them in terms of investment targets, value assessment, trading methods, risk and return, and professionalism.

　　First of all, I would like to talk about the macroeconomic impact on the stock market, real estate, funds, financial management, insurance. These five types of asset investment and macroeconomic rise or fall is pro-cyclical relationship. It is difficult to cross the economic cycle to achieve asset preservation and appreciation. When the macroeconomic situation is good and continues to rise, the investment prices and returns of these five types of assets usually continue to rise; when the macroeconomic situation is poor and continues to decline, the prices and returns of these five types of assets usually continue to fall. For example, the rise and fall of real estate and the stock market are barometers that reflect the rise and fall of the macroeconomic situation.

　　Secondly, I would like to talk about the relationship between luxury jewelry investment, gold investment and macro economy. Luxury jewelry investment and stock market, real estate, funds, wealth management, insurance investment logic is different, luxury jewelry investment through the economic cycle, through counter-cyclical adjustment to achieve asset preservation and appreciation. In the time of economic inflation, luxury jewelry investment can hedge inflation risk, stabilize asset prices and achieve appreciation. In the period of economic deflation, investment channels are reduced, and luxury jewelry provides high-quality investment targets, thereby ensuring and improving the return on assets. In addition, luxury jewelry investment is mainly circulated among high net value customers, and the entry threshold is high, which is not affected by the rise or fall of the mass income. In contrast, gold investment also has the function of preserving and increasing value, but the entry threshold of gold investment is low, affected by the rise or fall of mass income. Its cross-border liquidity is small, and it is obviously weaker than luxury jewelry in ensuring and improving the return on assets.

　　At present, the decline of stock prices, the decline of real estate prices, the decline of fund investment returns, and the contraction of financial management scale have significantly reduced the growth rate of household assets. In contrast, luxury jewelry investment is more preservation, convenience, low risk, high yield and other characteristics, can make assets keep flat and increase.

一、名贵珠宝投资的比较优势
The Comparative Advantage of Luxury Jewelry Investment

名贵珠宝投资与黄金投资、股市投资、债市投资、房地产投资、基金投资、理财、期货投资等，都是资产配置的重要组成部分。比较而言，名贵珠宝投资具有以下显著优势。体积小且价值高，稀有性，方便携带出国，维护成本低，国内外变现方便，资产安全，不记名，是富人把钱换一种方式存起来的最好方法，这些优势都是很多其他资产不具备的。投资名贵珠宝不是花钱，是用另外一种更保值的方式存钱。它唯一的风险就是你的钱没有买到相应价值的好珠宝，这就需要专业的名贵珠宝顾问为你把关。

Luxury jewelry investment and gold investment, stock market investment, bond investment, real estate investment, fund investment, wealth management, futures investment, etc., are all important components of asset allocation. In comparison, luxury jewelry investment has the following significant advantages: small size and high value, rarity, easy to carry abroad, low maintenance costs, easy to cash at home and abroad, safe assets, anonymous. These advantages making luxury jewelry the best way for the rich to save money in another way, and they not obtained by many other assets. Investing in luxury jewelry is not spending money, it is saving money in another way that is more stable. The only risk is that your money does not buy the appropriate value of good jewelry, which requires a professional jewelry consultant to double check for you.

红宝石戒指

Comparative Analysis of Luxury Jewelry Investment and Other Investments

In comparison, investment in precious gemstones has the characteristics of value preservation, convenience, low risk, and high returns, allowing residents to maintain and increase the value of their assets.

（一）独特的审美价值
Unique Aesthetic Value

名贵珠宝通常具有独特的颜色、切割和清晰度等特征，极具审美吸引力。这种独特性可以增加名贵珠宝的美感，同时提升名贵珠宝的市场价值。

Luxury jewelry often has unique characteristics such as color, cut and clarity, which are extremely aesthetically appealing. This uniqueness can increase the beauty of luxury jewelry and enhance the market value of luxury jewelry.

（二）易于保存
Easy to Store

名贵珠宝是一种实物投资，相比于其他投资方式更容易保存。通过保险箱、专业库房等专业化保管措施，可以确保名贵珠宝安全和完好无损。只要妥善保管，名贵珠宝价值就不会轻易消失，为投资者提供长期持有的信心和保障。

Luxury jewelry is a physical investment and is easier to preserve than other investment methods. Through professional storage measures such as safe deposit boxes and professional warehouses, luxury jewelry can be ensured safe and intact. As long as it is properly kept, the value of luxury jewelry will not easily disappear, providing investors with long-term confidence and protection.

小雨老师与斯里兰卡的最知名的矿主一起参加宝石展

（三）可携性和高流动性
Portability and high mobility

名贵珠宝通常紧凑而轻，体积小，易于携带。此外，名贵珠宝买卖是不记名的，稀有性和市场需求一直都很高，是全球范围内公认的珍贵资产，在全球范围内都有很高的流通性，方便投资者在不同市场间进行买卖，可选择国际拍卖行或者私人交易平台进行交易。这使得名贵珠宝投资更加灵活和便捷。

Luxury jewelry is usually compact and light, small in size, and easy to carry. In addition, the sale of luxury jewelry is anonymous, the rarity and market demand had always been high. It is recognized globally as a luxury asset, having a high liquidity in the worldwide range. It is convenient for investors to buy and sell between different markets, and also allow to choose international auction houses or private trading platforms for trading. This makes investment in luxury jewelry more flexible and convenient.

（四）资产保值增值
Maintaining and Increasing the Value of Assets

名贵珠宝，如红宝石、蓝宝石等，极具稀缺性，其价值通常较为稳定，不易受到经济波动影响。特别是在名贵珠宝供应短缺的情况下，增加的市场需求会推高名贵珠宝价格。随着时间的推移，名贵珠宝的价值通常会不断增长。在长期的经济周期中，名贵珠宝的价值始终保持着坚挺的走势，为投资者提供了稳定的回报。

Luxury jewelry, such as rubies, sapphires, etc., is extremely scarce, and its value is usually relatively stable and not susceptible to economic fluctuations. Especially in the case of a shortage of luxury jewelry supply, the increased market demand will push up the price of luxury jewelry. Over time, the value of luxury jewelry usually increases. In the long economic cycle, the value of luxury jewelry has always remained strong, providing investors with stable returns.

（五）对冲通货膨胀风险
Hedging Against Inflation Risks

名贵珠宝通常不受通胀的直接影响，因为名贵珠宝的价值与货币没有直接联系。在通胀环境下，名贵珠宝等有形资产是一种相对稳定的投资选择。

Luxury jewelry is usually not directly affected by inflation because the value of luxury jewelry is not directly linked to the currency. In an inflationary environment, tangible assets such as luxury jewelry are a relatively stable investment option.

小雨老师在纽约的名贵宝石故事分享
Ms. Xiaoyu Shared her Luxury Jewelry Story in New York

我在纽约的47街工作的时候，才了解到47街成为珠宝街的来历，犹太人在"二战"时期被纳粹迫害，他们的其他固定资产都被侵占，唯一带出来的是一些"藏在"鞋底、帽子里的名贵珠宝，悄悄地藏到纽约47街的一些石板下面，非常隐蔽，每次拿出一颗名贵珠宝来换一些生活物资。随着纳粹的倒台，犹太人重获新生，而47街慢慢就变成了名贵珠宝交易的地方，直到现在全球名贵珠宝交易最集中的地方依旧是纽约。大克拉的名贵珠宝，几乎都掌握在犹太人手里，几代相传，价值无限。这些犹太人主要服务全球的富豪以及各国皇室。

小雨老师在美国纽约47街的珠宝交易中心

When I was working on 47th Street in New York, I came to know how 47th Street became a jewelry street. Jews were persecuted by the Nazis during World War II, and their other fixed assets were seized. The only thing they brought out was some luxury jewelry hidden in the soles of shoes and hats, which were quietly hidden under some slabs on 47th Street in New York. They used one luxury piece of jewelry at a time in exchange for some living supplies. With the fall of the Nazis and the rebirth of the Jewish people, 47th Street slowly became a place for luxury jewelry trading, and until now the world's largest concentration of luxury jewelry trading is still New York. The great carat's luxury jewels, almost all in Jewish hands, have been passed down through generations and are of unlimited value. These Jews mainly serve the world's rich and royal families.

二、名贵珠宝投资与黄金投资的比较分析
Comparative Analysis of Luxury Jewelry Investment and Gold Investment

黄金投资是居民在短期和中期抵消通货膨胀的主要避险手段。俗话说，"乱世藏黄金"，受到地缘战争冲突、经济风险等不稳定因素的影响，传统投资渠道收益率普遍下降，使居民增强了对黄金保值增值属性的认识。例如，人们会将"小金豆"作为日常重要的储蓄方式。

Gold investment is the main hedge against inflation in the short and medium term. As the old saying goes, "hide gold in troubled times", affected by unstable factors such as geopolitical wars and conflicts and economic risks, the return rate of traditional investment channels had generally declined. This making the residents having enhanced their understanding of the value preservation and appreciation of gold. For example, people will use "little golden beans" as an important way to save money on a daily basis.

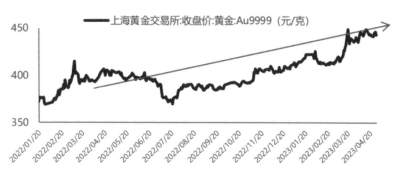

上海黄金交易所价格
Picture: Shanghai Gold Exchange prices
资料来源：上海黄金交易所，民生证券研究院
Source: Shanghai Gold Exchange, Minsheng Securities Research Institute

（一）黄金投资的不足之处
The Shortcomings of Gold Investment

黄金及其饰品的主要缺点，是产品溢价能力比较弱，缺乏资产增值的强劲空间。一是因为就饰品属性而言，国内黄金饰品品牌自身的溢价能力较弱；二是因为黄金本身的溢价能力较低。黄金材质的柔软性决定了其难以使用更为精美的加工工艺，市场上的黄金饰品设计感较弱、总体单一、同质化比较严重。黄金饰品价格通常和金价成正比关系，"按克计价"使价格更为透明，这使得黄金饰品价格总体趋近。如果考虑投资属性，选择金条作为投资品，其毛利率更是低于黄金饰品，因为金条是"按克计价"，而黄金饰品是"按件计价"。

因此，从长期收益视角来看，以黄金作为投资品获得的收益率相对较低，基本只能维持保值和微量增值。比较而言，名贵珠宝具有货值高、溢价高、容易携带的优点。

The main drawback of gold and its jewelry is that the product premium ability is relatively weak, and the lack of strong space for asset appreciation. First, in terms of jewelry attributes, the domestic gold jewelry brand's own premium ability is weak; Second, gold itself has a low premium. The softness of gold material determines that it is difficult to use more exquisite processing technology, and the design sense of gold jewelry on the market is weak, poor of differentiated products, and the homogenization is more serious. The price of gold jewelry is usually proportional to the price of gold, and "pricing by gram" makes the price more transparent, which makes the price of gold jewelry generally close. If you consider the investment attributes, the choice of gold bars as investment products, its gross profit margin is lower than gold jewelry, because gold bars are "per gram", and gold jewelry is "per piece".

Therefore, from the perspective of long-term income, the rate of return obtained by gold as an investment product is relatively low, and basically can only maintain value preservation and trace appreciation. In comparison, luxury jewelry has the advantages of high value, high premium and easy to carry.

（二）名贵珠宝投资和黄金投资在投资价值方面的区别
The Difference Between Luxury Jewelry Investment and Gold Investment in Terms of Investment Value

名贵珠宝的投资价值，主要取决于其品质、稀有性、市场需求和供应情况等因素。一颗高品质的名贵珠宝，如果其颜色、透明度、切工等方面都达到极佳的状态，那么它的投资价值就会非常高。此外，稀有的名贵珠宝，如钻石中的粉钻、红钻等，由于其稀缺性，价格往往也会高涨。市场需求也是影响名贵珠宝投资价值的重要因素。如果某一种类的名贵珠宝市场需求量大，而供应量相对较少，那么它的价格就会上涨，投资价值也随之提高。

The investment value of luxury jewelry mainly depends on factors such as its quality, rarity, market demand and supply. A high-quality luxury jewelry, if its color, transparency, cutting and other aspects have reached an excellent state, then its investment value will be very high. In addition, rare and valuable jewelry, such as pink diamonds, red diamonds, etc., Because of their scarcity, the price is often high. Market demand is also an important factor affecting the investment value of luxury jewelry. If there is a high demand for a certain type of luxury jewelry and a relatively low supply, its price will rise, and the value of the investment will increase.

黄金是一种传统的避险资产，被广泛认为是保值避险的首选之一。在经济不稳定、通货膨胀等情况下，人们往往会选择购买黄金来保值。因此，黄金的投资价值通常受到通货膨胀、政治不稳定和货币波动等因素的影响。此外，黄金的开采成本、全球供需情况等因素，也会对黄金的价格产生影响。

Gold is a traditional safe-haven asset and is widely regarded as one of the first choices for hedging. In the case of economic instability, inflation, etc., people often choose to buy gold to preserve their value. Therefore, the investment value of gold is often affected by factors such as inflation, political instability, and currency fluctuations. In addition, gold mining costs, global supply and demand conditions and other factors will also have an impact on the price of gold.

（三）名贵珠宝投资和黄金投资在用途方面的区别
The Difference Between Luxury Jewelry Investment and Gold Investment in Terms of Usage

名贵珠宝通常用于珠宝制造，也可以作为消费和投资收藏。名贵珠宝的价值主要体现在美学、稀缺性和金融属性等方面。

黄金不仅是一种贵金属，还被广泛用于珠宝加工、工业和科技等领域。因为黄金具有稳定的价值，也常被用作避险资产和储备资产。

Luxury jewelry is often used in jewelry manufacturing and can also be collected for consumption and investment. The value of luxury jewelry is mainly reflected in aesthetics, scarcity and financial attributes.

Gold is not only a precious metal, but also widely used in jewelry processing, industry and technology. Because gold has a stable value, it is also often used as safe havens and storage asset.

（四）名贵珠宝投资和黄金投资在流动性方面的区别
The Difference Between Luxury Jewelry Investment and Gold Investment in Liquidity

名贵珠宝的流通性，属于硬通货，而且流通起来的价值比较高，单颗珠宝的价格就可以过千万或者上亿，对于有需求的富人群体，是很好的等价物。

The circulation of luxury jewelry belongs to hard currency, and the value of circulation is relatively high, and the price of a single jewelry can exceed tens of millions or hundreds of millions, which is a good equivalent for the wealthy groups in demand.

相比之下，黄金属于泛流通、大众化流通，全球黄金市场的开放性和透明度也使得黄金的流通更为便利，但是每克的价值是恒定的，流通的总体价值不太高。

In contrast, gold belongs to the universal circulation, popular circulation, the openness and transparency of the global gold market also makes the circulation of gold more convenient, but the value per gram is constant, the overall value of circulation is not too high.

（五）名贵珠宝投资和黄金投资在价格波动方面的区别
The Difference Between Luxury Jewelry Investment and Gold Investment in Terms of Price Fluctuations

帕拉伊巴

价格波动也是投资者需要考虑的一个重要因素。名贵珠宝的价格波动取决于矿区的产量，"市场上永远不缺资金，但是缺值得购买的名贵珠宝"，这是我们行业的经典语录。因为名贵珠宝是稀缺资源，是大自然的馈赠，需要天时地利人和才能挖到一颗好的名贵珠宝。

黄金的价格波动，则通常受到全球政治经济形势、通货膨胀和货币政策等因素的影响。这些因素可能导致黄金价格的短期波动，但从长期角度来看，黄金通常能够保持其价值。

Price fluctuations is also an important factor for investors to consider. The price of luxury jewelry fluctuates depending on the output of the mining area, "there is never a shortage of funds in the market, but there is a shortage of luxury jewelry worth buying", which is a classic quote of our industry. That is because luxury jewelry is a scarce resource. It is a gift of nature, which needs the right time and place to dig a good luxury jewelry.

The fluctuation of gold price is usually affected by global political and economic situation, inflation and monetary policy. These factors can cause short-term volatility in the price of gold, but from a long-term perspective, gold is usually able to hold its value.

（六）名贵珠宝投资和黄金投资在投资风险方面的区别
The Difference Between Luxury Jewelry Investment and Gold Investment in Terms of Investment Risk

投资名贵珠宝可能存在一定的技术风险，因为鉴定和评估名贵珠宝需要专业知识和经验。如果投资者缺乏相关的知识和经验，可能会误判名贵珠宝的价值，从而增加投资风险。此外，市场需求和供应情况也可能导致价格波动，从而增加投资风险。

There may be some technical risks involved in investing in luxury jewelry, as the identification and evaluation of luxury jewelry requires specialized knowledge and experience. If investors lack relevant knowledge and experience, they may misjudge the value of luxury jewelry, thereby increasing investment risks. In addition, market demand and supply conditions can also lead to price fluctuations, thus increasing investment risks.

相比之下，黄金作为全球公认的避险资产，其价格通常相对稳定。因此，投资黄金的风险相对较低。

总的来说，名贵珠宝具有独特的投资价值和魅力。投资者在选择适合自己的投资品时，应该根据自己的投资目标、风险承受能力和个人喜好来做出决策。投资名贵珠宝，更能实现资产保值增值。

In contrast, gold is a globally recognized safe havens asset and its price is usually relatively stable. Therefore, the risk of investing in gold is relatively low.

In general, luxury jewelry has a unique investment value and charm. Investors should make decisions based on their own investment objectives, risk tolerance and personal preferences when choosing an investment product that suits them. Investment in luxury jewelry, it will be more likely to achieve asset preservation and appreciation.

三、名贵珠宝投资与股市投资的主要区别
The Main Differences Between Luxury Jewelry Investment and Stock Market Investment

名贵珠宝投资和股市投资是两种截然不同的投资形式。2023 年中国股市上证综指、沪深 300 和深圳成指分别回调了 3.7%、11.4% 和 13.5%，但是从整体估值看，当前股市仍处于历史低位，交易热度也处于历史低位。2023 年和 2021 年相比，国内股票累计缩水近 5.4 万亿元人民币。未来股市回暖要经历相对较长的时期，投资者较难从股市投资中获得较高资产收益。名贵珠宝投资与股市投资相比，有更多的实物和艺术价值，在收益方面也更具稳定性。

2023年以来中国股市市场整体走低
The overall Chinese Stock Market has Declined Since 2023
资料来源：中诚信国际研究院、Wind
Source: China Integrity International Research Institute, Wind

Luxury jewelry investment and stock market investment are two distinct forms of investment. In 2023, the Shanghai Composite Index, Shanghai Shenzhen 300 and Shenzhen Component Index of the Chinese stock market returned by 3.7%, 11.4% and 13.5% respectively. But from the overall valuation point of view, the current stock market is still at a historical low, and the trading heat is also at a historical low. Compared with 2023 and 2021, domestic stocks have shrunk by nearly 5.4 trillion yuan. The stock market recovery in the future will go through a relatively long period, and it is difficult for investors to obtain high asset returns from stock market investment. Compared with stock market investment, luxury jewelry investment has more physical and artistic value, and is more stable in terms of income.

具体而言，名贵珠宝和股市投资的主要区别如下。

Specifically, the key differences between precious gemstones and stock market investing are as follows.

（一）在性质和形式方面的区别
Differences in Nature and Form

名贵珠宝投资是一种实物资产投资，投资者购买实际的名贵珠宝，如蓝宝石、红宝石等，也属于稀缺资源的范畴。

股市投资则是投资于公司的股票或证券等金融资产。股市投资涉及购买和持有股票（股权），投资者通过购买公司的股票，从而获得公司的一部分所有权。

Luxury jewelry investment is a type of physical asset investment where investors buy actual luxury jewelry, such as sapphires, rubies, etc. It also falls into the category of scarce resources.

Stock market investing is investing in a company's financial assets such as shares or securities. Stock market investing involves buying and holding shares (equity), where investors acquire a portion of the ownership of a company by purchasing shares of the company.

（二）在流动性方面的区别
The Difference in Liquidity

名贵珠宝的流通性，属于硬通货，而且流通起来的价值比较高，单颗珠宝的价格就可以过千万元或者上亿元，对于有需求的富人群体，是很好的等价物。

股市投资通常具有较高的流动性，投资者可以通过证券市场，相对容易地买卖股票或其他金融资产。

The circulation of luxury jewelry belongs to hard currency, and the value of circulation is relatively high, and the price of a single jewelry can exceed ten million yuan or hundreds of millions of yuan, which is a good equivalent for the wealthy groups in demand.

Stock market investments usually have high liquidity, and investors can buy and sell stocks or other financial assets relatively easily through the securities market.

（三）在市场波动性方面的区别
Differences in Market Fuctuations

名贵珠宝市场波动性较小，因为名贵珠宝的产量是不确定的，所以很少存在大资本干预，资本不会去赚有不确定因素的钱，毕竟资金成本也很高。

股市通常具有较高的波动性，价格可能因公司业绩、宏观经济因素、市场预期等多种因素影响而波动。

The market for luxury jewelry is less volatile because the output of luxury jewelry is uncertain, so there is rarely major capital intervention, and capital will not make money with uncertain factors, after all, the cost of capital is also high.

The stock market usually has high fluctuations, and the price may fluctuate due to a variety of factors such as company performance, macroeconomic factors, and market expectations.

（四）在收益形式方面的区别
Differences in the Form of Income

名贵珠宝投资的回报，是用时间换价值，需要长期持有，3年及以上才会有丰厚的回报。

股票投资的回报，包括股息收益和股价升值，公司支付股息，为投资者提供现金流。

The return of luxury jewelry investment is to exchange time for value, and it needs to be held for a long time, and there will be a rich return for 3 years or more.

The return on stock investment, including dividend yield and stock price appreciation, and the company paying dividends, providing investors with cash flow.

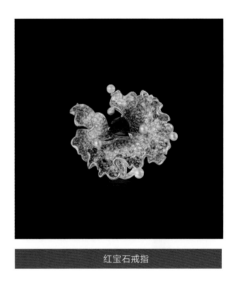

红宝石戒指

（五）在市场透明度方面的区别
Differences in Market Transparency

名贵珠宝市场在行业内透明，行业外相对不透明，鉴定和评估名贵珠宝的过程是非常专业的，准入门槛很高，为这个圈子构筑了很高的壁垒，这也是它能够安全又保值的优点，投资名贵珠宝不是花钱，是把钱用另外一种更保值的方式存钱。

The luxury jewelry market is transparent in the industry, relatively opaque outside the industry. The process of identification and evaluation of luxury jewelry is very professional, the entry threshold is very high. And it has built a high barrier for this circle, which is also the advantage of it, so it can be safe and value preservation. Investment in luxury jewelry is not to spend money, but to save money in another way that is more value preservation.

（六）在风险和保险方面的区别
Differences in Risk and Insurance

因为名贵珠宝较小，可能面临失窃、损坏或灾害的风险，但是从另外一个角度，它也是不容易被人发现的，且不记名，有利于隐藏自己的财富。

股市投资通常具有更高的风险和回报潜力，因为市场波动性更大，但同时也有更高的潜在收益。股市投资也涉及市场风险，但通常可以通过分散投资和其他风险管理策略来减轻一部分风险。

Because luxury jewelry is small, it may face the risk of theft, damage or disaster, but from another point of view, it is not easy to be found, and anonymous, which is conducive to hiding their wealth.

Stock market investments generally have a higher risk and return potential because the market is more volatile, but also have a higher potential return. Investing in the stock market also involves market risk, but it is often possible to mitigate some of that risk through diversification and other risk management strategies.

（七）在投资知识要求方面的区别
Differences in Investment Knowledge Requirements

鉴定名贵珠宝和了解名贵珠宝市场通常需要专业知识，投资者可能需要专业的名贵珠宝鉴定师的帮助。

股市投资也需要专业知识，但市场信息和研究通常更为透明和广泛。股市投资同样需要投资者对财务分析、公司业绩和市场趋势等方面有一定的了解。

Identifying luxury jewelry and understanding the luxury jewelry market often requires specialized knowledge, and investors may need the help of a professional luxury jewelry appraiser.

Investing in the stock market also requires expertise, but market information and research are generally more transparent and extensive. Stock market investment also requires an understanding of financial analysis, company performance and market trends.

四、名贵宝石投资与房地产投资的主要区别
The Main Differences Between Luxury Jewelry Investment and Real Estate Investment

小雨老师在泰国学习

从投资角度看，当前房地产的金融投资品属性极度弱化，房地产投资增速自2021年起急速下降，2023年和2021年相比，国内房地产资产缩水约6.8万亿元人民币。总体上看，伴随着人口结构变化和城镇化趋缓，房地产长期潜在需求可能见顶回落，房地产行业基本面仍然疲弱，仍需等待房地产需求端的复苏，房地产不再是居民财富的主要投资渠道。

From the investment point of view, the current property of real estate financial investment products is extremely weak. The growth rate of real estate investment has declined rapidly since 2021, and the domestic real estate assets have shrunk by about 6.8 trillion yuan in 2023 compared with 2021. In general, with the change of demographic structure and the slowdown of urbanization, the long-term potential demand for real estate may reach peak and fall. The fundamentals of the real estate industry are still weak, and it is still necessary to wait for the recovery of the real estate demand side, and real estate is no longer the main investment channel of residents' wealth.

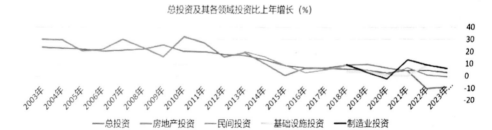

总投资及其各领域投资年增长率
Figure: Total Investment and Investment in Various Fields Increased by % Year-on-year
资料来源：中国财富管理50人论坛
Source: China Wealth Management 50 People Forum

具体而言，房地产不再是资产配置核心渠道的主要原因如下。一、从发展趋势看，中国潜在经济增速下滑，城市化速度减缓，这决定了房地产发展总体下降的长期趋势。二、从投资周期看，个别地方的房地产发展，前置了居民住房需求，资源错配促使房价在2017年后进入下行通道，损害了居民投资收益和投资信心。三、从当前市场来看，占据房企市场80%份额的民营房企处于业务收缩状态，占比房企市场20%份额的国有房企很难作为新开工主力军填补民营房企留下的空白市场，这很难促使房地产回暖。四、从供需关系看，人口红利拐点出现，长期人口变化使房地产供需关系发生实质性变化，房地产从供不应求变为供过于求。

Specifically, the main reasons why real estate is no longer the core channel for asset allocation are as follows. First, from the perspective of development trend, China's potential economic growth has declined and the urbanization rate has slowed down, which determines the long-term trend of the overall decline of real estate development. Second, from the perspective of the investment cycle, the real estate development in individual places has advanced the housing demand of residents, and the mismatch of resources has prompted the housing price to enter the downward channel after 2017, which has damaged the investment income and investment confidence of residents. Third, from the current market point of view, private housing enterprises, which occupy 80% share of the housing market, are in a state of business contraction. And state-owned housing enterprises, which account for 20% share of the housing market, are difficult to fill the blank market left by private housing enterprises as the main force of new construction, which is difficult to promote the real estate recovery. Fourth, from the perspective of supply and demand, the inflection point of demographic dividend has appeared. Long-term population changes have substantially changed the real estate supply and demand relationship, and the real estate has changed from short supply to oversupply.

因此，房地产市场持续在底部运行，房地产增速放缓降低了居民财产性收入增长预期，压缩了居民财富，找到具有稳定收益的投资渠道变得十分重要。

Therefore, the real estate market continues to run at the bottom, and the slowdown of real estate growth has reduced the growth expectation of residents' property income and compressed residents' wealth. It has become very important to find investment channels with stable returns.

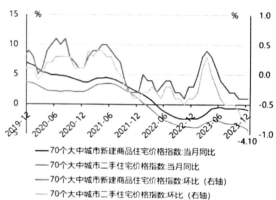

新房、二手房价格指数同比持续负增长
Total Investments and Their Annual Growth Rate in Various Sectors
资料来源：中诚信国际研究院、Wind
Source: China Wealth Management 50 People Forumear Growth

名贵珠宝作为存量时代的投资方式，与房地产具有明显的区别。
Luxury jewelry, as an investment method in the stock era, is obviously different from real estate.

（一）在投资对象方面的区别
Differences in Investment Objects

名贵珠宝投资，是可流动的实物资产投资，往往因其独特的稀有性、美观性和金融属性而备受追捧。

房地产投资，则主要关注的是不动产，即房产和地产，包括住宅、商业地产、工业地产等。

Luxury jewelry investment is a liquid real asset investment, often because of its unique rarity, aesthetic and financial attributes are highly sought after.

Real estate investment is mainly concerned with real estate, that is, housing estate and land estate, including residential, commercial real estate, industrial real estate and so on.

（二）在投资价值方面的区别
Differences in Investment Value

名贵珠宝的价值，主要取决于其稀有性、美观性因素。一些稀有的名贵珠宝，如钻石、红宝石、蓝宝石等，因其独特的品质和美感而具有极高的价值。

房地产的价值，则取决于地理位置、市场供需、房屋质量等多种因素。一个好的地理位置、市场需求大且供应不足的区域，以及高质量的房屋，都会使房地产价值上升。

The value of luxury jewelry mainly depends on its rarity and aesthetic factors. Some rare and luxury jewelry, such as diamonds, rubies, sapphires, etc., have a high value because of their unique quality and beauty.

The value of real estate depends on many factors such as geographical location, market supply and demand, and housing quality. A good location, an area with high market demand and low supply, and high quality homes could all make real estate values go up.

（三）在投资风险方面的区别
Differences in Investment Risks

由于名贵珠宝和房地产都是实物资产，因此都存在一定的风险。名贵珠宝投资的风险相对较小，因为名贵珠宝的每年的产量是不确定的，人为是没法干预产量的，再加上从以往的珠宝矿区产量经验来看，产量越来越少，而且名贵珠宝的品质和价值可以通过专业鉴定和评估来客观确定。

房地产投资的风险则相对较大，因为市场供需、经济环境、政策法规等多种因素都可能对房价产生影响。此外，房地产投资还需要考虑物业管理、维修保养等问题。

As luxury jewelry and real estate are both physical assets, there is a certain amount of risk. The risk of luxury jewelry investment is relatively small, because the annual production of luxury jewelry is uncertain, and human beings can not intervene in the production, coupled with the previous experience in the production of jewelry mining areas, the production is less and less, and the quality and value of luxury jewelry can be objectively determined through professional identification and evaluation.

The risk of real estate investment is relatively large, because market supply and demand, economic environment, policies and regulations and other factors may have an impact on housing prices. In addition, real estate investment also needs to consider property management, maintenance and other issues.

（四）在在流动性方面的区别
The Difference in Mobility

名贵珠宝投资的流动性属于特定流动，流动人群的对象有一定的门槛标准，交易通常需要通过专业的名贵珠宝市场或拍卖行进行。

房地产投资的流动性相对较高，因为房屋可以随时出租或出售。此外，房地产市场通常更加活跃，交易也更加频繁。房地产市场可能会有较大的波动性，受到经济周期、地区发展等多种因素的影响。

The mobility of luxury jewelry investment belongs to a specific flow, the object of the floating crowd has a certain threshold standard, and the transaction usually needs to be carried out through the professional luxury jewelry market or auction house.

Real estate investment is relatively liquid because houses can be rented or sold at any time. In addition, the real estate market is generally more active and transactions are more frequent. The real estate market can be highly volatile, affected by various factors such as economic cycles and regional developments.

（五）在投资回报方面的区别
Differences in Investment Returns

名贵珠宝是可以长期保值的资产，但具体的投资回报率会因市场状况和投资策略的不同而有所差异。在正常情况下，长期持有名贵珠宝，投资的回报率可能会高于房地产投资，主要是比较稀有，价格比较稳定，很少有政治干预。

相比之下，当前房地产供求关系发生重大变化，房地产投资不会保持以往的较高增长，与房地产投资相关的基础设施也相对饱和，很难获得较快增长，房地产在居民财富中的配置比例逐步下降。

Luxury jewelry is an asset that can hold its value for a long time, but the specific return on investment will vary depending on market conditions and investment strategies. Under normal circumstances, long-term ownership of luxury jewelry, the return on investment may be higher than real estate investment, mainly because it is relatively rare, the price is relatively stable, and there is little political interference.

In contrast, the current real estate supply and demand relationship has undergone major changes, real estate investment will not maintain a high growth in the past, the infrastructure related to real estate investment is relatively saturated. It is difficult to obtain rapid growth, and the allocation ratio of real estate in residents' wealth has gradually declined.

五、名贵珠宝投资与基金投资的主要区别
The Main Differences Between Luxury Jewelry Investment and Fund Investment

基金投资是资产配置的重要方式。当前，中国基金投资业务仍处于初步发展阶段，投资机构和投资者都需要不断提升资产配置能力和投研能力，全面的基金投顾服务体系仍需不断健全完善。受房地产发展下滑、地方债务规模扩大、中小银行问题、疫情等因素影响，基金投资收益自2021年7月以来逐年下降，基金也难以给投资者带来稳定增长的投资收益。未来居民资产配置的结构，既会压缩降低房地产等大类资产配置，也会充分考虑基金投顾业务的投资风险。

Fund investment is an important way of asset allocation. At present, China's fund investment business is still in the initial stage of development. Both investment institutions and investors need to continuously improve their asset allocation capabilities and investment research capabilities. And the comprehensive fund advisory service system still needs to be improved. Affected by factors such as the decline in real estate development, the expansion of local debt, the problems of small and medium-sized banks, and the epidemic, the investment income of the fund has declined year by year since July 2021. And the fund is also difficult to bring stable growth of investment income to investors. The structure of residents' asset allocation in the future will not only reduce the asset allocation of major categories such as real estate, but also fully consider the investment risk of fund investment consultant business.

在这种情况下，增加名贵珠宝等稳健增长型另类投资的投资比例，是确保资产保值增值的重要选择。

In this case, increasing the proportion of investment in stable growth alternative investments such as luxury jewelry is an important choice to ensure the preservation and appreciation of assets.

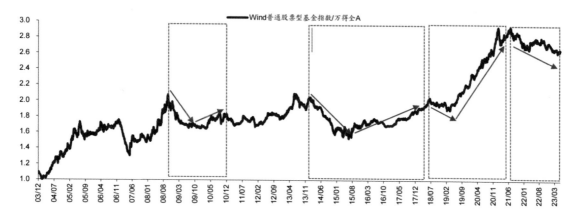

Wind普通股票型基金指数/万得全A
Wind Ordinary Stock Fund Index/Wind All A
资料来源：Wind，海通证券研究所
Source: Wind, Haitong Securities Research Institute

具体而言，名贵珠宝和基金投资的主要区别如下。
To be specific, the main differences between luxury jewelry and fund investment are as follows.

（一）在投资对象上的区别
The Differences in Investment Objects

名贵珠宝投资主要关注的是具有美观性和稀有性的珠宝。这些名贵珠宝，如钻石、红蓝宝石、祖母绿等，由于其稀缺性、美观性的因素，具有极高的投资价值。投资者往往会被其便利的流通性所吸引，从而选择进行名贵珠宝投资。这些名贵珠宝不仅具有独特的审美价值，还具有金融属性，能够实现保值和增值功能。

Luxury jewelry investment is mainly concerned with jewelry that is beautiful and rare. These precious jewels, such as diamonds, red sapphires, emeralds, etc., have a high investment value due to their scarcity and aesthetic factors. Investors are often attracted by its convenient liquidity and choose to invest in luxury jewelry. The luxury jewelry not only has unique aesthetic value, but also has financial attributes, which can realize the function of preserving and increasing value.

小雨老师与印度的矿主在一起交流

相比之下，基金投资是一种通过购买基金份额，将资金交由基金管理公司进行管理和运用的间接投资方式。基金投资的对象包括股票、债券等金融工具。投资者通过购买基金份额，将资金交给专业的基金管理公司进行管理。基金管理公司会根据市场情况，运用各种投资策略，将资金分散投资于不同的证券市场，以期实现投资的长期稳定增长。这种投资方式相对较为稳健，适合风险偏好较低的投资者。

In contrast, fund investment is a kind of indirect investment through the purchase of fund shares, the fund management company to manage and use the funds. The fund invests in financial instruments such as stocks and bonds. Investors buy fund shares and give their funds to professional fund management companies for management. Fund management companies will use a variety of investment strategies according to market conditions to diversify their funds into different securities markets in order to achieve long-term stable growth of investment. This investment method is relatively stable and suitable for investors with low risk appetite.

（二）在投资方式上的区别
The Differences in Investment Methods

名贵珠宝投资是一种直接投资于实物的行为。投资者购买并持有名贵珠宝本身，因此需要对名贵珠宝的品质、稀有性以及市场行情有一定的了解和判断。在自己不懂的情况下，需要专业名贵珠宝顾问的投资建议。

基金投资则是通过购买基金份额，间接投资于证券市场。投资者无需对具体的股票或债券进行判断，只需选择信任的基金管理公司并购买相应的基金份额即可。这种投资方式相对较为便捷，适合没有足够时间和专业知识的投资者。

Luxury jewelry investment is a kind of direct investment in the physical behavior. Investors buy and hold luxury jewelry itself, so they need to have a certain understanding and judgment of the quality, rarity and market conditions of luxury jewelry. If you do not understand the situation, you need the investment advice of a professional luxury jewelry consultant.

Fund investment is indirect investment in the securities market through the purchase of fund shares. Investors do not need to judge specific stocks or bonds, just choose a trusted fund management company and buy the corresponding fund shares. This type of investment is relatively convenient and suitable for investors who do not have enough time and expertise.

（三）在投资风险上的区别
The Differences in Investment Risk

名贵珠宝投资的最大风险，就是你的钱没有买到相应价值的珠宝。大多数人确实买到了高品质的名贵珠宝，但是花的价格太高了，很多珠宝商可能赚取了你未来三年的利润空间。但是如果你把时间维度拉长到10年来看，依然是很有投资价值的。所以我们一定要找专业的名贵珠宝资产配置顾问来帮你做更全面的珠宝筛选，通过支付服务费的形式，来避免大额利润的流失。

基金投资则通过分散投资于不同的证券，降低了个别证券的价格波动对整体投资的影响。即使某只股票或债券价格下跌，由于其他股票或债券的上涨，整体基金的投资收益仍可能保持稳定。这种投资方式相对较为稳定和安全。当前，人们在选择基金产品时，更加回归理性，不再追捧明星基金经理、新发基金和历史高收益基金。

小雨老师成为AIGS亚洲宝石学院特聘老师

The biggest risk of investing in luxury jewelry is that your money does not buy the appropriate value of jewelry. Most people do get high quality jewelry, but the price they spent are so high that many jewelers may have earned you a profit margin for the next three years. But if you extend the time dimension to 10 years, it is still a very valuable investment. Therefore, we must find a professional luxury jewelry asset allocation consultant to help you do a more comprehensive jewelry filtering, through the payment of service fees, to avoid the loss of large profits.

By diversifying investment into different securities, fund investment reduces the impact of price fluctuations of individual securities on the overall investment. Even if the price of a particular stock or bond falls, the investment return of the overall fund may remain stable due to the rise of other stocks or bonds. This kind of investment is relatively stable and safe. At present, people are more rational when choosing fund products, no longer chasing star fund managers, new development funds and historical high yield funds.

（四）在投资收益上的区别
The Differences in Investment Income

名贵珠宝需要通过时间来换价值，需要长期持有，再买进卖出，最终从一个小克拉的名贵珠宝换成一颗能上拍卖会的大克拉名贵珠宝。

基金投资的收益则取决于基金管理公司的管理和市场行情。虽然收益率可能不如名贵珠宝投资高，但相对较为稳定。此外，随着市场的发展和科技的进步，基金投资也更加便捷和多样化。投资者可以根据自己的需求和风险承受能力选择适合自己的基金产品。

Luxury jewelry needs to exchange for value through time. It requires to be held for a long time, then buy and sell, and eventually change from a small carat luxury jewelry to a large carat luxury jewelry that can be auctioned.

The return of the fund investment depends on the management of the fund management company and the market situation. Although the yield may not be as high as that of luxury jewelry investments, it is relatively stable. In addition, with the development of the market and the advancement of science and technology, fund investment is also more convenient and diversified. Investors can choose their own fund products according to their needs and risk tolerance.

六、名贵珠宝投资与理财投资的主要区别
The Main Differences between Luxury Jewelry Investment and Financial Investment

当前，银行理财市场规模收缩，呈现下降趋势，理财赎回余波仍存，2023年和2021年相比，国内银行理财累计缩水约2万亿元人民币。居民理财资产配置策略偏向保守，理财产品资产配置倾向于提高流动性，具有低风险、低波动性特征的固收类产品更受青睐，现金和银行存款配置占比提升。这意味着，居民投资偏好趋于保守，资产投资渠道减少，投资收益率降低。在这种情况下，名贵珠宝作为稳健增长型的投资方式，为居民资产保值增值提供了重要渠道。

At present, the scale of the bank financial market has shrunk, showing a downward trend. The aftermath of financial redemption still exists. Compared with 2023 and 2021, the cumulative shrinkage of domestic bank financial management was about 2 trillion yuan. Residents' financial asset allocation strategies tend to be conservative, and financial product asset allocation tends to improve liquidity. The fixed-income products with low risk and low volatility characteristics are more favored. Cash and bank deposit allocation proportion is showing an increasing trend. This means that residents' investment preferences tend to be conservative, asset investment channels are reduced, and investment returns are lower. In this case, luxury jewelry, as a steady growth type of investment, provides an important channel for the preservation and appreciation of residential assets.

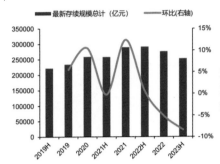

近3年中国全市场理财存续规模下降，H表示上半年
In the past three years, the scale of China's total market wealth management continued to decline. H represents the first half of the year

资料来源：银行业理财登记托管中心、信达证券研发中心
Source: Bank Financing Registration and Custody Center, Cinda Securities Research and Development Center

具体而言，名贵珠宝和理财投资的主要区别如下。
Specifically, the main differences between luxury jewelry and financial investment are as follows.

（一）在投资目标上的区别
Differences in Investment Objectives

名贵珠宝投资的目标是长期稳定化增长，这种投资方式需要投资者具备良好的现金流和对名贵珠宝的鉴赏能力，能够及时把握珠宝市场机遇。其实以我的经验来看，最难的不是卖珠宝，只要你的珠宝足够好，价格又相对比较有优势，肯定有人会接手，毕竟这是一个国际化的大市场，投资名贵珠宝的富人太多了。最难的是买珠宝，因为国外矿区的产量每年都在减少，这是大自然的产物，我们买到一颗价格合适的高品质珠宝真的太难了。你犹豫的那一下，可能珠宝矿主就卖给别人了。一般收到大珠宝的人，都不会着急马上卖，他们会放一段时间，3年，5年，甚至10年，一代人都有可能，这个会根据每个投资者的资金需求而定，在自己资金宽裕的情况下，越是把时间拉长，回报率越高。

理财的目标则更加多元化，它不仅关注投资回报，还注重财富的积累、保护和传承。理财需要综合考虑个人的收入、支出、资产和负债等因素，制定个性化的财务规划，实现长期的财务目标。

The goal of luxury jewelry investment is long-term stable growth, which requires investors to have good cash flow and appreciation of luxury jewelry, and be able to grasp the jewelry market opportunities in time. In fact, in my experience, the most difficult thing is not to sell jewelry, as long as your jewelry is good enough, the price is relatively advantageous. Surely someone will take over, after all, this is a big international market, and there are too many rich people investing in luxury jewelry. The most difficult thing is to buy jewelry, because the production of foreign mines is decreasing every year. It is a product of nature, and it is really difficult for us to buy a high-quality jewelry at the right price. The second you hesitated, the owner of the jewelry mine might have sold it to someone else. Generally, people who receive large jewelry will not worry about selling it immediately, they will put it for a period of time. 3 years, 5 years, or even 10 years, one generation will also be the chance. This will be determined according to the capital needs of each investor. As long as they hold enough money, the longer the time, the higher the rate of return.

小雨老师参加芭莎珠宝设计师展

The goal of wealth management is more diversified, focusing not only on investment returns, but also on the accumulation, protection and inheritance of wealth. Financial management requires comprehensive consideration of personal income, expenditure, assets and liabilities and other factors, to develop personalized financial planning, and to achieve long-term financial goals.

（二）在价值评估与波动性上的区别
The Differences between Value Assessment and Mobility

名贵珠宝的价值评估相对简单，可以参考每年的国际珠宝拍卖，比如佳士得、苏富比的拍卖。也可以去每年的国际珠宝展考察一下珠宝的市场行情

传统理财工具的价值也比较容易评估，并且市场波动性可能相对更可预测。传统理财工具的风险和回报，通常可以更容易地进行量化和管理。

The valuation of luxury jewelry is relatively simple and can be referred to the annual international jewelry auctions, such as Christie's and Sotheby's. You can also go to the annual international jewelry show to investigate the jewelry market.

The value of traditional financial instruments is also easier to assess, and market mobility can be relatively predictable. The risks and rewards of traditional financial instruments can often be more easily quantified and managed.

（三）在分析方法与依据上的区别
Differences in Analysis Methods and Basis

投资名贵珠宝需要对名贵珠宝市场和品质有深入了解的专业知识。名贵珠宝投资主要关注名贵珠宝市场的行情、价格走势、供求关系等因素，通过分析这些因素来预测市场走势，选择合适的投资时机和品种。

理财则更加注重个人的实际情况和需求，需要综合考虑个人的风险承受能力、投资期限、收益预期等因素，制定合适的投资策略和方案。

Investing in luxury jewelry requires in-depth knowledge of the luxury jewelry market and quality. Luxury jewelry investment mainly focuses on luxury jewelry market, price trend, supply and demand and other factors. By analyzing these factors to predict the market trend, choices can be made more wiser on the right investment time and variety.

On the other hand, financial management pays more attention to individual, actual situation and demand. It needs to comprehensively consider factors such as individual risk tolerance, investment period and income expectation to formulate appropriate investment strategies and programs.

（四）在投资范围上的区别
Differences in the Scope of Investment

名贵珠宝投资主要关注的是名贵珠宝这一特定资产的投资，包括名贵珠宝的购买、拍卖、租赁等交易活动。

理财则涵盖了更广泛的范围，包括投资、储蓄、保险、税务等多个方面，需要综合考虑各种财务工具和手段，实现整体的财务规划。

Luxury jewelry investment mainly focuses on the investment of luxury jewelry as a specific asset, including the purchase, auction, leasing and other trading activities of luxury jewelry.

Financial management covers a wider range, including investment, savings, insurance, tax and other aspects. It requires to consider a variety of financial tools and methods, to achieve the overall financial planning.

Differences in the Scope of Investment

七、名贵珠宝投资与投资理财型保险的主要区别
The Main Differences Between Luxury Jewelry Investment and Financial Insurance Investment

通常而言，投资理财型保险尤其是年金险，是高净值人群降低未来现金流风险的重要投资方式，以实现资产配置多元化、防范资金风险。但是，从整个保险行业来看，近年来保险行业负债端成本恶化将会导致资产端激进投资，从而形成恶性循环，保险超额收益率并未一直呈现上升趋势，而是呈现下降趋势。保险行业当前估值处于底部，资本市场波动较大，监管政策不断趋严。这些状况意味着，投资理财型保险并不能有效保障和提高资产收益率。相比之下，名贵珠宝投资更能穿越经济金融周期，监管政策更为友好，更能促进资产保值增值。

Generally speaking, financial insurance investment, especially annuity insurance, is an important investment way for high net value people to reduce future cash flow risks. In this way, they can achieve diversification of asset allocation and prevent capital risks. However, from the perspective of the entire insurance industry, in recent years, the deterioration of the liability side cost in the insurance industry will lead to aggressive investment in the asset side. As a result, a vicious circle will be formed. where the excess return of insurance has not always shown an upward trend, but a downward trend. The current valuation of the insurance industry is at the bottom. The capital market is volatile, and the regulatory policy is becoming stricter. These conditions mean that investment and financial insurance can not effectively protect and improve the return on assets. In contrast, luxury jewelry investment is more capable to pass through the economic and financial cycle. And the regulatory policy is more friendly, which can promote the preservation and appreciation of assets.

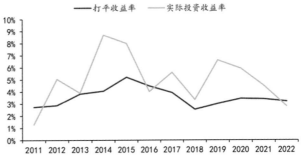

2011-2022年寿险行业成本与投资收益表
2011-2022 Life Insurance Industry Cost and Investment Income Statement
资料来源：王晴精算师，保险一哥，十三个精算师，方正证券研究院
Source: Wang Qing Actuary, Insurance Mr. No.1, 13 Actuaries, Founder Securities Research Institute

具体而言，名贵珠宝投资与投资理财型保险的主要区别如下。

To be specific, the main differences between luxury jewelry investment and investment financial insurance are as follows.

（一）在投资对象与风险特性上的区别
Differences in Investment Objects and Risk Characteristics

名贵珠宝投资关注的是具有独特魅力和稀有性的珠宝，具有良好的市场需求，溢价能力比较强，并且风险较低。相比之下，投资理财型保险是投资者购买保险合同，通过保险公司的专业投资来获取收益，虽然属于相对稳健的金融产品，但是保险资本市场波动较大，且监管约束较严，投资收益率较低。

Luxury jewelry investment focuses on jewelry with unique charm and rarity, which has good market demand, relatively strong premium ability, and low risk. In contrast, investment financial insurance means that investors purchase insurance contracts and obtain income through professional investment of insurance companies. Although it is a relatively stable financial product, the insurance capital market fluctuates greatly, and the supervision and constraint are strict, and the investment return rate is low.

（二）在投资回报上的区别
The Difference in Return on Investment

名贵珠宝投资的回报主要依赖于珠宝市场的涨跌和珠宝品质的提升，市场需求增加，名贵珠宝品质高且稀缺，投资者可能会获得高额的回报。相比之下，投资理财型保险的回报通常包括两部分：一部分是固定的保险利益，如身故保险金、生存保险金等；另一部分是投资收益，这部分收益取决于保险公司的投资表现。投资理财型保险收益偏稳健，但是溢价能力不高。

The difference in Return on Investment

The return on luxury jewelry investment mainly depends on the rise and fall of the jewelry market and the improvement of the quality of jewelry. With the increase of market demand, luxury jewelry is of high quality and scarce, and investors may get high returns. In contrast, the return of investment financial insurance usually includes two parts: one part is fixed insurance benefits, such as death insurance, survival insurance and so on; The other part is the investment income, which depends on the investment performance of the insurance company. The income of investment and financial insurance is stable, but the premium ability is not high.

（三）在流动性与变现能力上的区别
The Differences in Liquidity and Realization Ability

名贵珠宝的流动性通常较高，变现能力较强，投资者在需要资金时随时可以迅速将名贵珠宝变现。相比之下，投资理财型保险则通常具有较长投资时间期限，投资者可在到期时选择退保或贷款，流动性和变现能力相对较弱。此外，投资理财型保险更关注保障功能，如身故保险金等，只能在投资者身故后获得风险保障。

The liquidity of luxury jewelry is usually high, and the realization ability is strong, and investors can quickly cash in luxury jewelry at any time when they need funds. In contrast, investment financial insurance usually has a longer investment time period, investors can choose to surrender or loan at maturity. The liquidity and realization ability are relatively small. In addition, investment and financial insurance pays more attention to the protection function, such as death insurance, and can only obtain risk protection after the death of the investor.

（四）在投资期限上的区别
The Difference in the Investment Period

名贵珠宝投资通常没有固定的投资期限，投资者可以根据市场情况和自身需求随时调整投资策略，在高价时卖出从而获得较高投资收益。然而，投资理财型保险则通常具有固定的保险期限和投资期限，期限时间较长，投资者在届期时获得风险回报和相对较低的投资收益。

Luxury jewelry investment usually has no fixed investment period. The investors can adjust the investment strategy at any time according to the market situation and their own needs, and sell at high prices to obtain higher investment returns. However, investment and financial insurance usually has a fixed insurance period and investment period, and the period is longer. The investor can get the risk return and relatively low investment income at the expiration of the period.

Market Trends in Luxury Jewelry Investments

第三章
03 Chapter Three
名贵珠宝投资的市场趋势
The Market Trend of Luxury Jewelry Investment

一、全球主要名贵珠宝市场和交易中心
The World's Major Luxury Jewelry Market and Trading Center

　　总体上看，全球名贵珠宝市场正在逐步发展壮大，以下是全球主要名贵珠宝市场和交易中心的一些代表性地点。

　　In general, the global luxury jewelry market is gradually growing, and the following are some representative locations of the world's major luxury jewelry markets and trading centers.

蓝宝石

The World's Major Luxury Jewelry Market and Trading Center

In comparison, investment in precious gemstones has the characteristics of value preservation, convenience, low risk, and high returns, allowing residents to maintain and increase the value of their assets.

（一）印度孟买。印度是全球最大的名贵珠宝生产加工地。孟买被认为是世界上最大的名贵珠宝交易中心之一。这里有许多名贵珠宝店、制造商和批发商，涵盖了各种名贵珠宝，尤其是钻石和金饰品。在特征上，孟买的名贵珠宝贸易历史悠久，集聚了来自印度和其他国家的名贵珠宝商和买家。

Mumbai, India. India is the world's largest producer and processor of luxury jewelry. Mumbai is considered as one of the world's largest trading centers for luxury jewelry. There are many luxury jewelry shops, manufacturers and wholesalers covering all kinds of luxury jewelry, especially diamonds and gold jewelry. Characteristically, Mumbai's fine jewelry trade has a long history, bringing together fine jewelry and buyers from India and other countries.

（二）泰国曼谷。泰国是全球主要的红宝石和蓝宝石产地。曼谷是东南亚最大的名贵珠宝交易中心之一。这里汇聚了大量的名贵珠宝生产商、贸易商和加工商。在特征上，曼谷以其对彩色名贵珠宝的交易而闻名，包括蓝宝石、红宝石、翡翠等。

Bangkok, Thailand. Thailand is the world's leading producer of rubies and sapphires. Bangkok is one of Southeast Asia's largest trading centers for luxury jewelry. It is home to a large number of precious jewelry producers, traders and processors. Characteristically, Bangkok is known for its trade in colorful luxury jewelry, including sapphires, rubies, jadeites, etc.

（三）美国纽约。美国是全球主要的名贵珠宝进口、加工、消费和贸易国。纽约是全球最大的名贵珠宝市场之一，尤其以第47街的名贵珠宝区而闻名。这里有来自世界各地的名贵珠宝商和交易商。在特征上，纽约的名贵珠宝市场主要以钻石和彩色名贵珠宝为主，拥有大量的名贵珠宝展览和拍卖会。

New York, USA. The United States is the world's main jewelry importing, processing, consumption and trading country. New York is one of the largest markets for luxury jewelry in the world, especially known for its luxury jewelry section on 47th Street. There are famous jewelry and dealers from all over the world. In terms of characteristics, New York's luxury jewelry market is mainly dominated by diamonds and colored luxury jewelry, with a large number of luxury jewelry exhibitions and auctions.

（四）比利时安特卫普。安特卫普在钻石交易方面有着悠久的历史，是世界上最重要的钻石交易中心之一。在特征上，安特卫普的钻石交易主要集中在Diamond District（钻石区），这里有大量的钻石商、拍卖行和交易所。

Antwerp, Belgium. Antwerp has a long history in diamond trading and is one of the most important diamond trading centers in the world. In terms of characteristics, the diamond trade in Antwerp is mainly concentrated in the Diamond District, which has a large number of diamond dealers, auction houses and exchanges.

（五）中国内地和香港。中国内地名贵珠宝行业起步于20世纪80年代，是全球非常重要的名贵珠宝和玉石加工地之一。香港特区在亚洲地区是一个重要的名贵珠宝交易中心，吸引了大量的国际买家和卖家。在特征上，香港的名贵珠宝市场包括了各种类型的名贵珠宝，同时还是世界上最大的黄金市场之一。

Chinese Mainland and HK. The luxury jewelry industry started up since 1980's. It is one of the most important places of processing for luxury jewelry and jade in the world. The Hong Kong Special Administrative Region is an important trading center for luxury jewelry in Asia, attracting a large number of international buyers and sellers. In terms of characteristics, Hong Kong's luxury jewelry market includes all types of luxury jewelry, and it is also one of the largest gold markets in the world.

（六）瑞士日内瓦。日内瓦是瑞士著名的名贵珠宝和钟表交易中心，这里有一些高端的名贵珠宝店和拍卖行。在特征上，日内瓦以其高品质的蓝宝石、红宝石和其他贵重珠宝而著称。

Geneva, Switzerland. Geneva is Switzerland's famous luxury jewelry and watch trading center, there are some high-quality luxury jewelry shops and auction houses. Characteristically, Geneva is famous for its noble sapphires, rubies and other precious jewels.

（七）南非开普敦。南非是全球最大的钻石产地。开普敦在钻石采矿和交易领域具有重要地位，是全球最大的毛坯钻石交易中心之一。在特征上，这里有许多钻石加工和贸易公司，是南非著名的名贵珠宝产业中心。

Cape Town, South Africa. South Africa is the world's largest producer of diamonds. Cape Town has an important position in diamond mining and trading. It is one of the world's largest trading centers for rough diamonds. In terms of character, there are many diamond processing and trading companies here, and it is a well-known center of the luxury jewelry industry in South Africa.

（八）斯里兰卡科伦坡。斯里兰卡是全球著名的高端名贵珠宝产地，其蓝宝石储量占全球90%以上。科伦坡以蓝宝石和红宝石而闻名，是斯里兰卡名贵珠宝贸易的中心。

Colombo, Sri Lanka. Sri Lanka is the world's famous high-end jewelry source; its sapphire reserves account for more than 90% of the world. Colombo is famous for its sapphires and rubies and is the center of Sri Lanka's precious jewelry trade.

（九） 缅甸是全球最好的红宝石、蓝宝石、绝地尖晶石的生产大国。
Myanmar is the world's best producer of rubies, sapphires and Jedi spinel.

（十） 巴西是全球主要彩色名贵珠宝产地，比如亚历山大变石。
Brazil is the world's leading producer of colored luxury jewelry, such as Alexandrite.

（十一） 博茨瓦纳是全球最大的名贵珠宝级金刚石生产地。
Botswana is the world's largest producer of luxury jewelry-grade diamonds.

（十二） 俄罗斯名贵珠宝资源极为丰富，尤其是具有质量高、数量大的钻石资源。
Russia is extremely rich in luxury jewelry resources, especially with high quality and large quantity of diamond resources.

（十三） 日本是全球珍珠产量最大的国家。
Japan is the world's largest producer of pearls.

（十四） 澳大利亚是全球最好的黑欧泊生产大国、粉钻出产大国。
Australia is the world's best production country of black opal and pink diamond.

（十五） 坦桑尼亚是马亨盖尖晶石的产地。
Tanzania is the country of production of Mahenge spinel.

（十六） 哥伦比亚是祖母绿出产大国。
Colombia is a major emerald producer.

（十七） 莫桑比克是红宝石出产大国。
Mozambique is a major producer of rubies.

小雨老师与斯里兰卡的矿主交流

二、当前国内名贵珠宝市场特点
The Current Domestic Luxury Jewelry Market Characteristics

最近几年，中国名贵珠宝消费市场呈现出反弹上升的趋势。在短期内，中国名贵珠宝消费市场波动较大。随着疫情结束、经济回暖，高端消费回流等因素对名贵珠宝消费的需求上升比较快，中国名贵珠宝市场将会回暖并增长。此外，相较于发达国家和发达地区，中国人均名贵珠宝消费偏少，名贵珠宝消费市场仍有较大拓展开发空间，名贵珠宝品类结构进一步改善，名贵珠宝市场的盈利增值空间也比较充足。因此，从中长期看，中国名贵珠宝市场极具稳健性和成长性特征，名贵珠宝投资会获得优良的配置收益。

小雨老师与哥伦比亚祖母绿的商人交流

In recent years, China's luxury jewelry consumption market has shown a rebound trend. In the short term, China's luxury jewelry consumption market fluctuates greatly. With the end of the epidemic, the economic recovery, together with the return of high-end consumption and other factors, the demand for luxury jewelry consumption rises relatively fast. Thus, China's luxury jewelry market will pick up and grow. In addition, compared with developed countries and developed regions, China's per capita consumption of luxury jewelry is less, luxury jewelry consumer market still has a large space for expansion and development. Luxury jewelry category structure is further improved, and the luxury jewelry market profit appreciation space is also more sufficient. Therefore, in the medium and long term, China's luxury jewelry market has its stable and growing characteristics, and luxury jewelry investment will obtain excellent allocation returns.

在2020年上半年至2023年上半年的中国珠宝市场上，名贵珠宝是消费者和投资者重点关注的资产。尤其值得关注的是彩色名贵珠宝。2023年以来，彩色名贵珠宝市场销售火爆、量价齐升。据调研，2023年上半年中国全品类彩色名贵珠宝价格平均涨幅在30%—50%，大克拉或相对稀有的名贵珠宝价格涨幅则高达100%—150%。红宝石、蓝宝石等贵重珠宝平均克拉单价上涨约30%，其他品类如芬达石、海蓝宝石、摩根石等平均克拉单价上涨在10%—30%。（资料来源：中国名贵珠宝玉石首饰行业协会）

In the Chinese jewelry market from the first half of 2020 to the first half of 2023, luxury jewelry was the asset that consumers and investors focus on. Of particular concern is the colored jewelry. Since 2023, the color luxury jewelry market had been booming, and the volume and price have risen. According to the survey, in the first half of 2023, the average price increase rate of the whole category of colored luxury jewelry in China was 30%-50%, and the price increase rate of large carat or relatively rare luxury jewelry were as high as 100%-150%. The average carat unit price of luxury jewelry such as ruby and sapphire had risen by about 30%, and the average carat unit price of other categories such as Fanta stone, aquamarine and Morgan stone had risen by 10%-30%. (Source: China Luxury Jewelry Jade Accessary Industry Association)

三、全球和区域市场特点
Global and Regional Market Characteristics

（一）名贵珠宝投资的全球市场特征
Characteristics of the Global Market for Luxury Jewelry Investment

随着中国经济的发展，中国市场的增长潜力越来越受到重视。在中国大陆高端奢侈品市场上，蒂芙尼、卡地亚、梵克雅宝、施华洛世奇、潘多拉等国际著名品牌的市场占比也很高。在中国内地的中高端名贵珠宝市场上，具有极强影响力的是港资著名品牌，例如周大福、周生生等；以及内地著名品牌，例如老凤祥、潮宏基等。周大福在2017年4月拍出的"粉红之星"钻石（The CTF PinkStar Diamond），重达59.60克拉，以7120万美元成为全球拍卖会上成交的最昂贵天然钻石。

帕拉伊巴戒指

国内外著名品牌在布局中国重点市场的同时，也布局了中国跨区域市场。当前，东南亚经济增速高于全球平均水平，东南亚乐观的消费市场前景，也吸引高端名贵珠宝品牌积极布局。

With the development of China's economy, more and more attention has been paid to the growth potential of the Chinese market. In Chinese Mainland's high-end luxury product market, Tiffany, Cartier, Van Cleef & Arpels, Swarovski, Pandora and other international famous brands also have a high market share. In the high-end jewelry market of Chinese Mainland, Hong Kong-funded famous brands have a strong influence, such as Zhou Dafu, Zhou Shengsheng, etc. And some mainland's local famous brand, such as Lao Fengxiang, Chao Hongji, etc. The CTF Pink Star Diamond, a 59.60-carat diamond sold by Chow Tai Fook in April 2017, was the most expensive natural diamond ever sold at auction in the world at $71.2 million.

These famous home and abroad brands not only allocating their markets in China as the key markets, but also in China's cross-regional market. At present, the economic growth rate in Southeast Asia is higher than the global average, and the optimistic consumer market prospects in Southeast Asia also attract high-end and luxury jewelry brands to actively layout.

（二）名贵珠宝的品质和稀缺性仍是重点
The Quality and Scarcity of Luxury Jewelry is Still the Focus

名贵珠宝的稀缺性和供应量对其价值产生重大影响。虽然有新矿藏的开发，但是出产量还是非常低，所以市场上对名贵珠宝一直处于供不应求的状态。

The scarcity and availability of luxury jewelry has a significant impact on its value. Although there is the development of new mineral deposits, but the production is still very low, so the market for luxury jewelry has been in short supply.

小雨老师拜访北京地质大学的郭颖院长

虽然技术的不断发展可能改变名贵珠宝市场，例如，合成名贵珠宝的生产技术不断进步，可能影响天然名贵珠宝的市场需求和价值。但是，高品质、稀有和特殊的名贵珠宝可能会更受欢迎。对于少见的、高品质名贵珠宝的需求可能会持续增加，这些名贵珠宝在市场上的稀缺性可能会提高其价值。

Although the continuous development of technology may change the luxury jewelry market, for example, the continuous advancement of the production technology of synthetic luxury jewelry may affect the market demand and value of natural luxury jewelry. However, high quality, rare and special jewelry may be more popular. The demand for rare, high-quality jewelry is likely to continue to increase, and the scarcity of these jewelry in the market is likely to increase their value.

（三）名贵珠宝投资的资产准入提升
The Asset Access of Luxury Jewelry Investment is Improved

名贵珠宝投资属于高品质生活的重要组成部分，也是资产配置多元化、防范风险、提升资产收益率的重要方式，因此资产准入门槛也相对较高。根据胡润研究院的研究，家庭净资产达 600 万元人民币的属于富裕阶层，家庭净资产达1000万元人民币的属于高净值阶层，净资产达1亿元人民币的属于超高净值阶层，家庭净资产达 3000万美元的属于国际超高净值阶层。

总体而言，第一，家庭净资产在3000万元人民币以下的家庭，在投资规划上更接近普通人群思维习惯，更关注家庭日常消费等常规项目，风险承受能力相对较低，更关注现金流。因此，对于这部分家庭而言，包括名贵珠宝在内的高品质生活属于柔性支出，配置的比例相对较低，可以关注投资型的名贵珠宝，适当降低名贵珠宝的配置资金，或者缩短名贵珠宝的持有时间，及时卖出，从而提升名贵珠宝的保值和变现能力。

第二，家庭资产在3000万元人民币以上的家庭，更加追求包括名贵珠宝在内的高品质生活、奢侈消费和家族财富传承，有足够的高端消费需求和传承眼光，现金流和变现相对不太重要，风险承受能力较高。因此，对于这部分家庭而言，包括名贵珠宝在内的高品质生活属于必要支出，配置比例相对较高，可以关注投机型名贵珠宝，适当提高名贵珠宝的配置资金，或者延长名贵珠宝的持有时间以获得长期较高投资收益，从而提升名贵珠宝的投资能力、抗风险能力和传承能力。

Luxury jewelry investment is an important part of high-quality life, but also an important way to diversify asset allocation, prevent risks, and improve asset returns, so the asset access threshold is relatively high. According to the research of Hurun Research Institute, those with a family net value of 6 million yuan belong to the rich class, those with a family net value of 10 million yuan belong to the high net value class, those with a family net value of 100 million yuan belong to the ultra-high net value class, and those with a family net value of 30 million dollars belong to the international ultra-high net value class.

In general, firstly, families with a net value of less than 30 million yuan are closer to the thinking habits of ordinary people in investment planning. They are paying more attention to routine items such as daily consumption, having a relatively low risk tolerance, and paying more attention to cash flow. Therefore, for this part of the family, including luxury jewelry, high-quality life is a flexible expenditure, so the proportion of allocation is relatively low. You can pay attention to the investment type of luxury jewelry, appropriately reduce the funds allocation of luxury jewelry, or shorten the holding time of luxury jewelry, with timely sale, so as to improve the value of luxury jewelry and liquidity.

Secondly, families with family assets of more than 30 million yuan are more in pursuit of high-quality life, luxury consumption and family wealth inheritance, including luxury jewelry. They are having enough high-end consumption demand and inheritance vision. Their needs for cash flow and realization are relatively less important, and their risk tolerance is high. Therefore, for these families, high-quality life including luxury jewelry is a necessary expenditure, and the allocation ratio is relatively high. They can pay attention to speculative and luxury jewelry, appropriately increase the funds allocation of luxury jewelry, or extend the holding time of luxury jewelry to obtain long-term higher investment returns. In this case, the investment ability, anti-risk ability and inheritance ability of luxury jewelry can be improved.

四、可持续名贵珠宝投资的重要性
The Importance of Sustainable Jewelry Investment

在名贵珠宝投资中，要有长期可持续的投资理念。在经济变局年代，资产红利在变化和转型，资产配置的投研思维和投研框架需要全面革新。要树立长期主义理念，既要考虑投资成本和利润，更要关注投资安全，充分考虑各种非市场因素，树立投资的底线思维和极限思维。要打破传统投资中短视化的确定性稀缺，关注资产的成长性稀缺，对房地产等传统投资"做减法"，对名贵珠宝等另类资产"做加法"。资产配置的核心理念是管控风险，实现资产长期保值增值。名贵珠宝投资，正是资产长期保值增值的重要投资渠道。

In the investment of luxury jewelry, we must have a long-term sustainable investment concept. In the era of economic changes, asset dividends are changing and transforming, and the investment and research thouhgt and framework of asset allocation need to be comprehensively reformed. To establish a long-term philosophy, we should not only consider investment costs and profits, but also pay attention to investment safety, fully consider various non-market factors, and establish bottom-line thought and limit thinking in investment. It is necessary to break the short-sighted certainty scarcity in traditional investment, and pay attention to the growth scarcity of assets. "Subtract" from traditional investments such as real estate, and "Add" to alternative assets such as luxury jewelry. The core concept of asset allocation is to control risks and realize long-term asset preservation and appreciation. Luxury jewelry investment is an important investment channel for long-term asset preservation and appreciation.

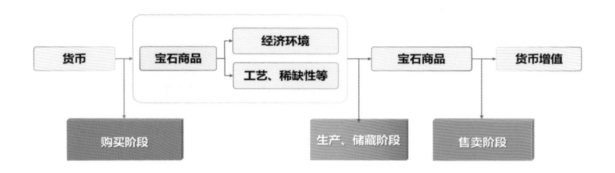

名贵珠宝投资及其增值过程
Luxury Jewelry Investment and Its Appreciation Process
资料来源：笔者自制
Source: Self-made by the Author

首先，名贵珠宝投资具有保值增值的功能，可以确保资产配置风险可控，实现资产的风险管控目标。

其次，在资产配置中选择名贵珠宝投资，将货币转化为名贵珠宝的内在价值，再附加名贵珠宝长期看涨的市场行情，以及对名贵珠宝的稀缺性和创意设计，能够让名贵珠宝在后期售卖阶段获得较高的资产收益率。

再次，名贵珠宝的购买投资和资产增值是循环递增的过程。刚开始的时候，较小的货币资本只能购买投资价值较小的名贵珠宝，即较小的货币资本转化为较小的名贵珠宝商品资本。但是，名贵珠宝未来增值的附加值，能够让刚开始的较小货币资本获得较高的投资收益，并再次购买投资价值更高的名贵珠宝，让更多的货币资本转化为更大的名贵珠宝商品资本。依次循环，最终通过名贵珠宝的投资，实现资产的保值增值。

First of all, luxury jewelry investment has the function of preserving and increasing value, which can ensure that the risk of asset allocation is controllable, and achieve the target of asset risk control.

Secondly, the selection of luxury jewelry investment in asset allocation is converting currency into the intrinsic value of luxury jewelry. Coupled with the long-term bullish market conditions of luxury jewelry, as well as the scarcity and creative design of luxury jewelry, we can make luxury jewelry obtain a higher return on assets in the later stage of sale.

Thirdly, the purchase and investment of luxury jewelry and the appreciation of assets are cyclic and increasing processes. At the beginning, the smaller monetary capital can only buy luxury jewelry with a smaller investment value, in other words, the smaller monetary capital is converted into a smaller luxury jewelry commodity capital. However, the added value of the luxury jewelry in the future can enable the small monetary capital in starting point to obtain a higher investment income. And buy the luxury jewelry with a higher investment income again, so that more monetary capital is transformed into a larger luxury jewelry commodity capital. Cycle in turn, and finally you can achieve asset preservation and appreciation through the investment of luxury jewelry.

此外，可持续名贵珠宝投资还要关注名贵珠宝投资的中期变现分红逻辑。客户和名贵珠宝咨询服务商签订合作协议，有两种合作模式，第一种，由名贵珠宝咨询服务商以客户资金协助客户挑选购买名贵珠宝，所有权归客户，经营权也归客户，客户需要缴纳一定的咨询服务费。第二种情况，所有权归客户，经营权归名贵珠宝咨询服务商，一般咨询服务协议为3年起，因为一颗高品质的名贵珠宝一般在3年后才会出来新的市场价格。名贵珠宝咨询服务商在市场上寻找合适的交易机会帮助客户卖出，如果使客户获得较高的投资收益，名贵珠宝咨询服务商可收取一定比例的管理费和一定比例的名贵珠宝溢价收益，如果跌价卖出或者在3年以内客户自己卖出，那名贵珠宝咨询服务商只收取管理费。

In addition, sustainable luxury jewelry investment should also pay attention to the medium-term realization dividend logic of luxury jewelry investment. The client and the luxury jewelry consulting service provider sign a cooperation agreement. There are two cooperation modes. In the first case, the luxury jewelry consulting service provider helps the client to select and purchase luxury jewelry with the client's funds. In the second case, the ownership belongs to the customer, and the management right is in the luxury jewelry consulting service provider, and the general consulting service agreement is 3 years or longer, which is because a high-quality luxury jewelry will generally come out with the new market price after 3 years. Luxury jewelry consulting service providers find the right trading opportunities in the market to help customers to sell. If the customer obtains a high investment income, the luxury jewelry consulting service providers can charge a certain percentage of management fees and a certain percentage of premium income of luxury jewelry. If the price is sold low, or the customer sell by himself within 3 years, the luxury jewelry consulting service providers only charge management fees.

五、案例研究：名贵珠宝投资案例
Case Study: Luxury Jewelry Investment Case

（一）名人和收藏家的名贵珠宝收藏
The Collection of Luxury Jewelry by Celebrities and Collectors

1.中国香港富商刘銮雄于2015年以1.8亿元人民币的高价拍下一颗重达16克拉的稀世粉钻，又以3亿元人民币拍下一颗名为"Blue Moon"的12克拉蓝钻。

Joseph Lau, a wealthy businessman from Hong Kong, China, bought a rare 16-carat pink diamond at a high price of 180 million yuan in 2015, and a 12-carat blue diamond named "Blue Moon" at a high price of 300 million yuan.

2.周大福于2017年以约合4.52亿元人民币拍下重达59.60克拉的"粉红之星"粉色钻石。

Chow Tai Fook bought the "Pink Star" pink diamond weighing 59.60 carats for about 452 million yuan in 2017.

3.赫伯特·巴菲特，以其庞大的名贵珠宝收藏而著称，其中包括蓝宝石、祖母绿、红宝石等。

Herbert Buffett, known for his vast collection of luxury jewelry, including sapphires, emeralds, rubies, etc.

4.陈岚，名贵珠宝资深投资者，于2013年在香港以8600万港元拍下一枚重达75.36克拉的水滴形钻石。她非常关注名贵珠宝的投资价值，配置了很多高品质大克拉的红宝石、蓝宝石、帕拉伊巴等名贵珠宝。

Chen Lan, a veteran investor in fine jewelry, paid HK $86 million for a 75.36-carat teardrop diamond in Hong Kong in 2013. She is very concerned about the investment value of luxury jewelry, and has configured a lot of high-quality large carat ruby, sapphire, Paraiba and other luxury jewelry.

5.刘嘉玲，也是资深的名贵珠宝投资者，她在投资珠宝上的眼光十分独到，拥有国际著名珠宝设计师Cindy Chao的作品，以及法国尚美的古董皇冠，这些稀有的珠宝每年都在升值。

Carina Lau is also a senior investor in luxury jewelry. She has a very unique vision in investing in jewelry. She owns the works of the internationally famous jewelry designer Cindy Chao, as well as the antique crown of the French beauty. These rare jewelry's value are increasing year by year.

小雨老师与演员刘嘉玲

（二）成功的名贵珠宝投资案例
Successful Investment Cases of Luxury Jewelry

1.蒂芙尼拥有一颗128克拉的黄钻，这颗宝石在1877年被发现，1878年，被以1.8万美元的价格买进。2019年，这颗黄钻的价值已经达到3000万美元，140年增长了1600倍的价值。

Tiffany owns a 128-carat yellow diamond that was discovered in 1877 and being trade with a price of $18,000 in 1878. In 2019, the value of this yellow diamond has reached $30 million, which is 1600 times increase in value in 140 years.

2.秀兰·邓波儿的父亲于1940年，邓波儿12岁生日的时候，以7210美元的价格买入一颗9.54克拉的深蓝色钻石。2016年3月18日，美国纽约苏富比拍卖行对这颗蓝钻的估价已经达到了惊人的2500万美元至3500万美元。

Shirley Temple's father bought a 9.54 carat dark blue diamond for $7,210 on Temple's 12th birthday in 1940. On March 18, 2016, Sotheby's auction house in New York estimated that the blue diamond had reached a staggering $25 million to $35 million.

小雨老师总结
Summary by Ms. Xiaoyu

宝石品类	买入年份	买入价格（美元）	卖出（年份）	卖出价格（美元）	持有年数	翻倍
Shirley Temple蓝钻	1940	7210	2016	22,000,000	76	3051
Blue Belle of Asia蓝宝石	1926	65000	2014	17,300,000	88	266
Hancock红钻	1956	13500	1987	880,000	31	65
Hancock紫粉钻和红紫钻	1956	6500	1987	200,000	31	31
La Peregrina珍珠项链	1969	37000	2011	2,368,500	42	64
Sunrise Ruby（以Gerhad的买入成本计算）	1980	200000	2015	30,000,000	35	150

以上是一些富商早期购买的名贵珠宝，佩戴了几年又卖出获得的收益。成功的名贵珠宝投资者通常具备深厚的专业知识、市场敏感度和价值认知。名贵珠宝在二次交易的时候，大多数情况下主要价值来源于主石，而不是镶嵌设计。名贵珠宝即便佩戴了很多年，只要没有损坏主石，也不太会影响价格，所以名贵珠宝只会越放越保值。

The above is some of the early purchase of valuable jewelry by wealthy businessmen. They wore a few years and sell for returns. Successful luxury jewelry investors often have deep expertise, market sensitivity and value perception. In the second transaction of luxury jewelry, in most cases the main value comes from the main stone, rather than the Mosaic design. Even if the luxury jewelry being worn for many years, as long as there is no damage to the main stone, it will not affect the price, so luxury jewelry will only be more and more valuable.

（三）名贵珠宝投资失败的案例分析
Case Analysis of Luxury Jewelry Investment Failure

1.20世纪初，美国总统泰迪·罗斯福投资了一颗名贵珠宝，相信其价值巨大。然而，后来发现这颗名贵珠宝的品质并不如他最初认为的那样高，导致了投资失败。

In the early 20th century, US President Teddy Roosevelt invested in a precious jewel, believing it to be of great value. However, it was later discovered that the quality of this precious jewel was not as high as he initially thought, leading to the failure of the investment.

2.英国富商尼克斯·尼古拉斯曾投资一颗巨大的钻石，但后来发现其实际价值远低于购买价格。他因此遭受了巨大的财务损失。

The wealthy British businessman Nicklas Nicholas once invested in a huge diamond, but later found that its actual value was far less than the purchase price. He suffered huge financial losses as a result.

这些案例都有一个共同点，他们对名贵珠宝品质的错误估计、把握不准市场价值导致的投资决策错误。我们在投资名贵珠宝，一定要寻求专业的鉴定和名贵珠宝资产配置顾问的建议，以减少投资风险。

These cases all have one thing in common, they misestimate the quality of luxury jewelry, grasp the wrong market value and led to the wrong investment decisions. When we invest in luxury jewelry, we must seek professional appraisal and luxury jewelry asset allocation consultant's advice to reduce investment risk.

（四）名贵珠宝投资失败的教训与分析
Lessons and Analysis of Luxury Jewelry Investment Failure

小雨老师经验分享
Ms. Xiaoyu's experience Sharing

我给大家总结一下大多数人投资名贵珠宝失败的原因：
1.缺乏专业的名贵珠宝鉴赏知识。
2.珠宝未获得专业鉴定和证书。
3.名贵珠宝买入价格太高，导致短期内变现亏钱。
4.名贵珠宝投资也需要良好的风险管理策略，包括分散投资，定期评估投资组合，了解个人的风险承受能力。名贵珠宝投资不是单纯买一颗，是需要搭配组合的，有的宝石是可以短期变现的，有的宝石是需要长期持有才获得增值的，但是往往很多人并没有持续的稳定现金流，并没有好的规划珠宝的组合配置。可以参考我书中写的关于名贵珠宝风险管理的部分。
5.未寻求专业名贵珠宝评估师或名贵珠宝资产配置顾问，导致投资决策不明智。名贵珠宝投资是一个特殊领域，专业知识对于成功投资至关重要。

Let me summarize the reasons why most people fail to invest in luxury jewelry:
1. Lack of professional appreciation knowledge of luxury jewelry.
2. The jewelry has not obtained professional appraisal and certificate
3. The purchase price of luxury jewelry is too high, resulting in short-term loss of money with realiztion.
4. Luxury jewelry investment also requires a good risk management strategy, including diversification, regular evaluation of the portfolio, and understanding of the individual's risk tolerance. Luxury jewelry investment is not just to buy one piece, ti needs to be combined. Some jewelries can be realized in the short term, and some jewelries need to be held for a long time to gain value. But often many people do not have continuous stable cash flow, and there is no good planning of the combination of jewelries. You can refer to the section on risk management of luxury jewelry as written in my book.
5. Did not asking a professional luxury jewelry appraiser or luxury jewelry asset allocation consultant for help, had resulted in unwise investment decisions. Luxury jewelry investing is a special area where expertise is essential for successful investing.

Supply and Demand, Price and Value of Luxury Jewelry

第四章
04 Chapter Four
名贵珠宝的供需关系、价格与价值
The Supply and Demand Relationship, Price and Value of Luxury Jewelry

一、名贵珠宝的保值与增值潜力
The Preservation and Appreciation Potential of Luxury Jewelry

（一）决定名贵珠宝保值增值潜力的必要条件
The Necessary Conditions for Determining the Value Preservation and Appreciation Potential of Luxury Jewelry

名贵珠宝保值增值，最重要的是满足两个条件：一个是漂亮，一个是稀有。这两个条件必须同时满足，它才会有保值增值的可能。围绕这两个主要条件，我们来分析一下其他影响因子，我们会有具体每一个品种的名贵珠宝的价值量化标准，大家可以仔细品鉴。

To preserve or increase luxury jewelry's value, the most important thing is to meet two conditions: one is beautifulness, one is rarity. These two conditions must be met at the same time, and then it will have the possibility of preserving and increasing value. Around these two main conditions, let's analyze other impact factors, and we will have a specific value quantification standard of each variety of luxury jewelry, so that you can carefully appreciate.

尖晶石戒指

The World's Major Luxury Jewelry Market and Trading Center

The most important thing for precious gems to maintain and increase their value is to meet two conditions: one is beautiful and the other is rare. two conditions must be met at the same time before it can maintain or increase its value.

1.漂亮：从三个维度来量化一颗名贵珠宝的漂亮
Beautifulness: from Three Dimensions to Quantify the Beauty of a Luxury Jewelry

第一，颜色。这是指名贵珠宝的色彩饱和度高。比如蓝宝石的皇家蓝跟矢车菊蓝，红宝石的鸽血红，帕拉伊巴的霓虹色，帕帕拉恰的日出色、日落色等，这些都是名贵珠宝很稀有的颜色。颜色饱和度越高价值越高。

First, color. This refers to the high color saturation of luxury jewelry. For example, the royal blue of sapphire and cornflower blue, the pigeon blood of ruby, the neon color of Paraiba, the outstanding and sunset color of Papalacha, etc. These are very rare colors of luxury jewelry. The higher the color saturation, the higher the value.

第二，净度。这是指名贵珠宝的干净程度。它主要分为玻璃体全干净的，以及有一点点瑕疵的，有一点点矿缺的，有一点点色带，或者金红石针的。一颗名贵珠宝越大，我们对净度的要求应该越低。一颗1克拉的名贵珠宝，我们要求它95%干净。一颗10克拉的名贵珠宝，我们要求它80%干净就可以了。名贵珠宝净度越好，价格越高。

Second, clarity. This refers to the cleanliness of valuable jewelry. It is mainly divided into vitreous that are completely clean, and those with a little blemish, a little ore deficiency, a little ribbon, or rutile needle. The larger a precious jewel, the lower our requirements for clarity should be. For a one carat jewel, we want it to be 95% clean. For a 10-carat jewel, we want it to be 80% clean. The clearer the jewelry, the higher the price.

第三，切工。指的是名贵珠宝打磨的工整度、抛光度。也就是说，名贵珠宝放在你面前，它是否一眼看上去很闪亮。切工主要注意两个地方：一个是空窗效应，就是台面过大，厚度很薄，就会形成中间的那部分没有颜色或者颜色很浅。第二个是切割过厚，台面会显得很小，它的厚度很厚会有点发暗不舒展。名贵珠宝切割越亮越舒展，价格越高。

Third, cut. This refers to the degree of fineness and polishing of luxury jewelry. That is to say, when the luxury jewelry is placed in front of you, whether it looks very shiny at first glance. The cutter mainly pays attention to two places: one is the empty window effect, that is, the platform is too large, the thickness is very thin, and the middle part will be formed without color or the color is very light. The second is that the cutting is too thick, the platform will appear very small, its thickness is very thick will be a little dark and not stretch. The brighter the cut, the more stretch, the higher the price.

2.稀有：从三个维度来量化一颗名贵珠宝的稀有
Rarity: Quantify the Rarity of a Precious Gem from Three Dimensions

红宝石戒指

第一，产地。也就是名贵珠宝出产的地方，一颗名贵珠宝的产地很影响它整体的价值。以蓝宝石为例，条件都相同的情况下，已经绝矿了的克什米尔产区的蓝宝石价格是远远超过缅甸产区和斯里兰卡产区的价格的。一颗缅甸抹谷产区的高品质红宝石，它的价格也是要远远高于一颗莫桑比克红宝石的价格的。知名矿区或者绝矿的矿区名贵珠宝价格高于普通矿区。

First, origin. That is, the place where luxury jewelry is produced. The origin of a luxury jewelry affects its overall value. In the case of sapphire, for example, under the same conditions, the price of sapphire in the Kashmir producing area is far higher than that in Myanmar and Sri Lanka producing areas. The price of a high-quality ruby from Mogok, Myanmar, is also much higher than the price of a Mozambique ruby. The price of luxury jewelry in well-known mining areas or exceptional mining areas is higher than that in ordinary mining areas.

第二，重量。名贵珠宝是有克拉溢价的，比如1到3克拉的名贵珠宝，它们的差距并不大，比如1克拉价格是5万元，2克拉可能是10万元，3克拉可能是30万元。但如果这个名贵珠宝变成了5克拉跟10克拉，那可能5克拉就要100万元以上，10克拉可能就要1000万元以上。因为名贵珠宝越大，还能保证它净度好，颜色好，非常难得，价值肯定越高。

小雨老师教学工作照

Second, weight. Luxury jewelry is a carat premium. For example, for 1 to 3 carat luxury jewelry, their difference is not large, where 1 carat price is 50,000 yuan, 2 carat may be 100,000 yuan, 3 carat may be 300,000 yuan. But if this luxury jewelry becomes 5 carats and 10 carats, it may be more than 1 million yuan for 5 carats and more than 10 million yuan for 10 carats. Because the larger the luxury jewelry, and its clarity guarantee to be so good, and the color is so good. Tt is very rare, and for sure the higher the value must be.

第三，优化。这是指这颗名贵珠宝是天然的还是人工后期干预过的，其实主要是分两种情况，如果是加热，比如说红宝石的加热，蓝宝石的加热，国际上是认可的，仍然是非常有价值的。在美国，大多数蓝宝石都是加热的，但价格依然很贵。但是帕帕拉恰、尖晶石的加热在国际市场上接受度并不高，一旦加热过，价格就会很低。

Third, optimization. This refers to whether this luxury jewelry is natural or artificial later intervention. In fact, it is mainly divided into two cases. If it is heating, such as ruby heating, sapphire heating, is recognized internationally, it is still very valuable. In the United States, most sapphires are heated, but they are still expensive. However, the heating of Papalacha and spinel is not highly accepted in the international market, and once heated, the price will be very low.

(二) 影响名贵珠宝保值增值潜力的重要因素
Important Factors Affecting the Value Preservation and Appreciation Potential of Luxury Jewelry

名贵珠宝的保值及增值潜力，受到多种因素的影响。在购买名贵珠宝时，要咨询专业的名贵珠宝鉴定师意见，确保购买到高品质、具有保值增值潜力的名贵珠宝。同时，也要关注市场动态和消费者需求的变化，以便更好地把握名贵珠宝的投资和收藏价值。

The value preservation and appreciation potential of luxury jewelry is affected by many factors. When purchasing luxury jewelry, it is necessary to consult a professional luxury jewelry appraiser to ensure the purchase of high-quality luxury jewelry with the potential to preserve and increase value. At the same time, it is also necessary to pay attention to changes in market dynamics and consumer demand in order to better grasp the investment and collection value of luxury jewelry.

1.名贵珠宝保值增值潜力受名贵珠宝品质的影响
The Value Preservation and Appreciation Potential of Luxury Jewelry is Affected by the Quality of Luxury Jewelry

优质名贵宝石通常具有鲜艳的颜色、高透明度及精美的切工，这些特质使得名贵宝石在市场上更受欢迎，从而具有更高的市场价值及保值增值潜力。例如，一颗颜色鲜艳、透明度高的钻石，往往比一颗颜色暗淡、透明度低的钻石更具有保值增值潜力。

此外，储藏保养对名贵宝石的品质影响也非常大。妥善的保养和维护能够保持名贵宝石的品质，从而更好地保持其价值，并增强其保值及增值潜力。例如，一颗经过妥善保养的钻石，其光泽和透明度能够保持得更好，从而具有更高的市场价值及保值增值潜力。

High-quality luxury jewelry usually has bright colors, high transparency and exquisite cuts, which make luxury jewelry more popular in the market, and thus having a higher market value and the potential for preservation and appreciation. For example, a diamond with a bright color and high transparency often has more potential to preserve and increase value than a diamond with a dim color and low transparency.

In addition, storage and maintenance on the quality of luxury jewelry is also very large. Proper care and maintenance can maintain the quality of precious jewelry, thereby better preserving its value and enhancing its preservation and appreciation potential. For example, a diamond that has been properly maintained can maintain its shine and transparency better, thus having a higher market value and potential for preservation and appreciation.

2.名贵珠宝保值增值潜力受名贵珠宝稀缺性的影响
The Value Preservation and Appreciation Potential of Luxury Jewelry is Affected by the Scarcity of Luxury Jewelry

某些特定种类、特定产地的名贵珠宝由于其稀缺性，往往具有更高的市场价值及保值增值潜力。例如，产自某些特定矿山的红宝石或蓝宝石，其独特的成分和纹理使得它们在市场上更受欢迎，从而具有更高的市场价值及保值增值潜力。

名贵珠宝的稀缺性提升了知名度和价值。对于商品而言，其定价能力决定了该商品能否获得更高的价值。对于普通商品，没有长期价值和定价能力，其利润主要是渠道利润，经常是以降低流通成本来提升流通环节利润。但是，名贵珠宝由于其稀缺性，在历史沉淀中更能够形成长期品牌价值，具有更强的定价能力。名贵珠宝在产品端和情感、身份和自我价值方面相结合，以特定文化内涵提升定价能力。名贵珠宝具有稀缺性，在渠道端能够占据主动地位、把握主动权，在渠道优势中提升名贵珠宝品牌的知名度，提升名贵珠宝价值。

Because of its scarcity, some luxury jewelry of specific types and specific origin often has higher market value and potential for preservation and appreciation. For example, rubies or sapphires produced from certain mines, due to their unique composition and texture, make them more popular in the market, and thus have a higher market value and potential for preservation and appreciation.

The scarcity of luxury jewelry increases visibility and value. For a commodity, its pricing power determines whether the commodity can obtain a higher value. For ordinary commodities, there is no long-term value and pricing power, and their profits are mainly channel profits, which are often improved by reducing circulation costs. However, due to its scarcity, luxury jewelry is more able to form long-term brand value in the historical precipitation and has stronger pricing power. Luxury jewelry combines emotion, identity and self-worth at the product end to enhance pricing power with specific cultural connotations. Luxury jewelry is scarce, and it can occupy an active position and grasp the initiative in the channel side, so as to enhance the brand awareness of luxury jewelry and enhance the value of luxury jewelry in the channel advantages.

3.名贵珠宝保值增值潜力受消费需求的影响
The Value Preservation and Appreciation Potential of Luxury Jewelry is Affected by Consumer Demand

一是名贵珠宝保值增值潜力受市场需求影响。当某种名贵珠宝在市场上受到追捧时，其价格往往会上涨，从而进一步增强其保值及增值潜力。例如，近年来，随着消费者关注度的提高，绿色翡翠的市场需求增加，其价格也相应上涨，从而进一步增强了其保值及增值潜力。

First, the value preservation and appreciation potential of luxury jewelry is affected by market demand. When a certain luxury jewelry is sought after in the market, its price tends to rise, further enhancing its preservation and appreciation potential. For example, in recent years, with the increase of consumer attention, the market demand for green jadeite has increased, and its price has also risen accordingly, further enhancing its preservation and appreciation potential.

二是名贵珠宝保值增值潜力受投资需求的影响。在投资需求中，确保财产安全和财产增值居于首位。财产安全在生活中非常重要，因此人们会把名贵珠宝作为投资手段，实现资产保值增值功能，并作为财富的象征。从历史和现实看，凡是在社会动荡、局势不稳的时期，名贵珠宝都以其保值增值功能备受青睐。

Second, the value preservation and appreciation potential of luxury jewelry is affected by investment demand. Among the investment needs, ensuring the safety of property and the appreciation of property are at the top. Property security is very important in life, so people will use luxury jewelry as an investment means to achieve the function of asset preservation and appreciation, and as a symbol of wealth. From the point of view of history and reality, in the period of social unrest and instability, luxury jewelry is favored for its value preservation and appreciation function.

三是名贵珠宝的保值增值潜力集中体现在名贵珠宝的符号价值中。名贵珠宝在所有商品中属于特殊商品，实际使用价值并非名贵珠宝的核心价值，因为名贵珠宝并不能直接满足人们的生理需求。名贵珠宝的真正价值是符号价值，是以名贵珠宝自身的稀缺性，满足人们情感归属需求、身份尊重需求和自我价值需求。

Third, the preservation and appreciation potential of luxury jewelry is concentrated in the symbol value of luxury jewelry. Luxury jewelry is a special commodity in all commodities, the actual use value is not the core value of luxury jewelry, because luxury jewelry can not directly meet people's physiological needs. The real value of luxury jewelry is symbolic value, which is based on the scarcity of luxury jewelry itself to meet people's needs of emotional belonging, identity respect and self-value.

四是名贵珠宝的保值增值潜力受美好生活需求影响。名贵珠宝的保值增值功能受情感因素、身份尊重需要和自我价值的影响。名贵珠宝通常象征着爱情，婚礼首饰、订婚钻戒也通常将名贵珠宝作为必备消费品。名贵珠宝通常象征着一定的身份、地位和财富。名贵珠宝还象征着持有者的深厚文化底蕴和文化内涵。这种爱情消费观、身份消费观和自我价值消费观和投资观，大大提升了名贵珠宝的价值。

Fourth, the value preservation and appreciation potential of luxury jewelry is affected by the demand for a better life. The function of preserving and increasing value of luxury jewelry is affected by emotional factors, the need of identity respect and self-value. Expensive jewelry is usually a symbol of love, wedding jewelry, engagement rings are also often expensive jewelry as a must-have consumer goods. Luxury jewelry often symbolizes a certain identity, social status and wealth. Luxury jewelry also symbolizes the holder's profound cultural heritage and cultural connotation. This concept of love consumption, identity consumption, self-value consumption and investment view greatly enhance the value of luxury jewelry.

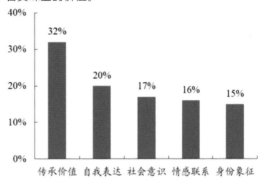

我国消费者对黄金和珠宝存在多样化需求
Chinese Consumers Have Diversified Demand for Golden Luxury Jewelry
资料来源：世界黄金协会、开源证券研究所
Source: World Gold Council, Open Source Securities Institute

二、名贵珠宝供需动态及其对价格的影响
The Supply and Demand Dynamics of Luxury Jewelry and Its Impact on Prices

名贵珠宝市场的供需动态是一个复杂而多变的现象，它受到多种因素的影响，包括矿产资源的分布、开采活动的规模、市场供需的变化等。这些因素共同作用，决定了名贵珠宝市场的价格走势。

The dynamic of supply and demand of luxury jewelry market is a complex and changeable phenomenon, which is affected by many factors, including the distribution of mineral resources, the scale of mining activities, and the change of market supply and demand. These factors work together to determine the price trend of the luxury jewelry market.

（一）市场需求对名贵珠宝价格的影响
The Impact of Market Demand on the Price of Luxury Jewelry

随着中国经济社会不断发展，以及消费观念的转变升级，名贵珠宝消费场景也日益变化，名贵珠宝日常佩戴和投资需求不断增长，名贵珠宝消费量逐步提升。人们的悦己需求，大大拓展了名贵珠宝的投资市场空间。

With the continuous development of China's economy and society, as well as the transformation and upgrading of consumption concepts, the consumption scene of luxury jewelry is also changing day by day, the daily wear and investment demand of luxury jewelry is growing, and the consumption of luxury jewelry is gradually increasing. People's demand for pleasure has greatly expanded the investment market space of luxury jewelry.

cindychao珠宝展的受邀嘉宾

（二）名贵珠宝资源短缺对名贵珠宝价格的影响
The Impact of the Shortage of Luxury Jewelry Resources on the Price of Luxury Jewelry

从供给端来看，名贵珠宝的矿产分布对名贵珠宝价格的影响也比较大。名贵珠宝市场的资源供需动态较为复杂，由多种因素共同决定，包括矿产资源的分布、开采活动的规模、市场供需的变化等。这些因素共同作用，左右了名贵珠宝市场的价格走势。

名贵珠宝的供应，主要取决于名贵珠宝矿产资源的分布，以及开采活动的规模。一些稀有或珍贵的名贵珠宝，如红宝石、蓝宝石等，其供应受到地理分布的限制，开采难度大，成本高，因此供应量相对较少。名贵珠宝的供应量一直短缺，而市场对名贵珠宝的投资需求却一直在增加，名贵珠宝价格往往会上涨。这是因为供需关系失衡，供不应求导致价格不断攀升。

From the supply side, the mineral distribution of luxury jewelry has a relatively large impact on the price of luxury jewelry. The dynamic of resource supply and demand of luxury jewelry market is more complex, which is jointly determined by many factors, including the distribution of mineral resources, the scale of mining activities, and the change of market supply and demand. These factors work together to influence the price trend of the luxury jewelry market.

The supply of luxury jewelry mainly depends on the distribution of luxury jewelry mineral resources and the scale of mining activities. Some rare or precious luxury jewelry, such as rubies, sapphires, etc., their supply is limited by geographical distribution, mining is difficult, and the cost is high, so the supply is relatively small. The supply of luxury jewelry has been in short supply, while the market's investment demand for luxury jewelry has been increasing, and the price of luxury jewelry tends to rise. This is because the relationship between supply and demand is out of balance, and the shortage of supply leads to rising prices.

小雨老师在四川美术学院任教

（三）消费者偏好对名贵珠宝价格的影响
The Impact of Consumer Preferences on the Price of Luxury Jewelry

此外，消费者偏好的变化也会引起名贵珠宝价格的波动。例如，如果某种名贵珠宝突然变得非常受欢迎，需求可能会激增，导致价格上涨。这是因为需求增加而供应有限，价格上涨是市场供需平衡的结果。相反，如果某种名贵珠宝被发现具有严重的缺陷或问题，可能导致需求下降，价格下跌。这是因为消费者对这种名贵珠宝的信心降低，需求减少导致价格下跌。

需要注意的是，名贵珠宝市场的价格受到多种因素的影响，包括但不限于供需关系、市场供需、经济环境、政策法规等。因此，要准确预测名贵珠宝市场的价格走势，需要综合考虑各种因素，并进行深入的市场研究和分析。同时，投资者和消费者也需要关注市场动态和趋势变化，以便做出明智的决策和购买决策。

帕帕拉恰戒指

In addition, changes in consumer preferences can also cause fluctuations in the price of luxury jewelry. For example, if a certain kind of luxury jewelry suddenly becomes very popular, the demand may surge, causing prices to rise. This is because demand is increasing and supply is limited, and the price rise is the result of the balance between supply and demand in the market. On the contrary, if a certain luxury jewelry is found to have serious defects or problems, it may lead to a drop in demand and a drop in price. This is because consumers have less confidence in this luxury jewelry, and reduced demand has led to lower prices.

It should be noted that the price of the luxury jewelry market is affected by a variety of factors, including but not limited to supply and demand relation, market supply and demand, economic environment, policies and regulations. Therefore, to accurately predict the price trend of the luxury jewelry market, it is necessary to consider various factors comprehensively and carry out in-depth market research and analysis. At the same time, investors and consumers also need to pay attention to market dynamics and trend changes in order to make informed decisions and purchase decisions.

（四）名贵珠宝资产配置表
Luxury Jewelry Asset Allocation Table

名贵珠宝资产配置主要是根据标准普尔家庭资产配置建议表来的，名贵珠宝的配置属于积极投资，占到家庭资产的30%，需要通过专业的知识才能保证资产增值。我个人建议想要投资名贵珠宝，可以选择自己专门去学习，也可以选择咨询名贵珠宝投资顾问，将自己的风险降低。

The asset allocation of luxury jewelry is mainly based on the Standard & Poor's household Asset allocation recommendation table. The allocation of luxury jewelry is an active investment, accounting for 30% of household assets, and professional knowledge is required to ensure the appreciation of assets. I personally suggest that if you want to invest in luxury jewelry, you can choose to learn it by yourself, or you can choose to consult a luxury jewelry investment consultant to reduce your risk.

第一，名贵珠宝价格区间与投资人总资产和投资人稳定流动之间的关系，名贵珠宝投资除了专业的宝石鉴赏能力，还需要持续的流动资金，因为名贵珠宝是一个需要长时间持有的资产，正常情况下，时间越长收益越大。拥有稳定的流动资金，可以第一时间买到品质好、价格好的名贵珠宝，如果错过时间点，就需要花更高的价格去买进，就不划算。另外，名贵珠宝投资不是配置一颗宝石，而是多颗不同价位宝石的组合资产。购买珠宝是一个随机的行为，稳定的流动资产可以随时购买预算范围内的珠宝。我整理了一组数据说明三者之间的关系（仅供参考），总资产情况：600万元以下，600万元—1,000万元，1,000万元—3,000万元，3,000万元—1亿元，1亿元以上。稳定的流动资产：50万元—100万元，100万元—500万元，500万元—1,000万元，1,000万元—3,000万元，3,000万元—5,000万元，5,000万元—1亿元，1亿元以上。珠宝价格区间：30万元—50万元，50万元—100万元，100万元—500万元，500万元—1,000万元，1,000万元—3,000万元，3,000万元—5,000万元，5,000万元—1亿元，1亿元以上。

名贵珠宝资产配置表的两种配置模型
Two Kinds of Allocation Models of Luxury Jewelry Asset Allocation Table

First, the relationship between the price range of luxury jewelry and the total assets and the stable flow of the investors. In addition to professional gem appreciation, luxury jewelry investment also needs continuous working capital, as luxury jewelry is an asset that needs to be held for a long time. Under normal circumstances, the longer the time, the greater the return. With stable liquidity, you can buy high-quality and expensive jewelry at the first time, if you miss the time point, you need to spend a higher price to buy. Ad that will be not cost-effective. In addition, luxury jewelry investment is not an allocation of a stone, but a combination of several stones at different prices. Buying jewelry is a random act, and stable liquid assets can be readily purchased within a budget. I have compiled a data to illustrate the relationship between the three (for reference only), and the total assets: less than 6 million yuan, 6 million to 10 million yuan, 10 million to 30 million yuan, 30 million to 100 million yuan, and more than 100 million yuan. Stable flexible assets: 500,000 to 1 million yuan, 1 million to 5 million yuan, 5 million to 10 million yuan, 10 million to 30 million yuan, 30 million to 50 million yuan, 50 million to 100 million yuan or more. Jewelry price range: 300,000 to 500,000 yuan, 500,000 to 1 million yuan, 1 million to 5 million yuan, 5 million to 10 million yuan, 10 million to 30 million yuan, 30 million to 50 million yuan, 50 million to 100 million yuan, more than 100 million yuan.

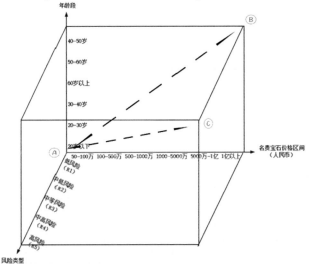

名贵珠宝价格区间与投资人抗风险能力的关系
The Relationship Between the Price Range of Precious Jewelry and Investors' Risk Resistance Ability

第二，名贵珠宝价格区间与投资人抗风险能力和投资人年龄之间的关系，投资人在20岁至30岁抗风险能力相对较低，40岁至50岁抗风险能力最强，根据我目前接触的投资案例来讲，40岁至50岁确实也是配置名贵珠宝的主力人群。根据20岁至60岁以上的人群划分，以及抗风险能力，还有名贵珠宝价格区间，请参考以下表格（仅供参考）。

Second, the relationship between the price range of luxury jewelry and the risk resistance and the age of the investors. The risk resistance of investors between 20 and 30 years old is relatively low, and the risk resistance of investors between 40 and 50 years old is the strongest. According to the investment cases I have been in contact with, 40 to 50 years old are indeed the main group of people who equip with luxury jewelry. Please refer to the following table (for reference only) which is grouped base on the age of people between 20 and over 60 years old, together with the resistance to risk, and the price range of luxury jewelry.

三、名贵珠宝价格趋势、市场动态与投资时机
Luxury Jewelry Price Trends, Market Dynamics and Investment Opportunities

（一）名贵珠宝的价格趋势
The Price Trend of Luxury Jewelry

相对于其他行业，名贵珠宝投资的行业天花板较高。名贵珠宝不同于其他普通商品，具有较高的单位价值量，通俗点说，就是名贵珠宝可以卖到较高的价钱。从市场分布看，名贵珠宝具有非常多的适用场景，能够满足婚恋、送礼、悦己等许多场景需求。这形成了较为庞大的市场规模和消费群体，赋予名贵珠宝消费较高的市场容量，使名贵珠宝价格稳步攀升。

Compared to other industries, the industry ceiling of luxury jewelry investment is higher. Luxury jewelry is different from other ordinary commodities, with a higher unit value, in popular terms, luxury jewelry can be sold at a higher price. From the perspective of market distribution, luxury jewelry has a lot of application scenarios, which can meet the needs of many scenes such as marriage, gifting, and pleasing oneself. This has formed a relatively large market size and consumer groups, giving a high market capacity for luxury jewelry consumption, so that the price of luxury jewelry has risen steadily.

名贵珠宝的价格变化趋势，通常受到市场波动的影响。名贵珠宝不同于一般商品的特征在于，名贵珠宝并非必须消费品，而是可选择性消费品。名贵珠宝作为可选择性商品，深受居民收入水平高低的影响。当社会经济恢复正常并增长较快的时候，大众收入水平增长较快，对名贵珠宝配置的需求就会增加。当经济低迷的时候，高净值资产家庭出于资产保值增值的目标，配置名贵珠宝的需求会上升。

The price trend of luxury jewelry is usually affected by market fluctuations. The characteristic of luxury jewelry is different from general commodities in that luxury jewelry is not a necessary consumer product, but an optional consumer product. As an optional commodity, luxury jewelry is deeply affected by the level of residents' income. When the social economy returns to normal and grows rapidly, the mass income level grows rapidly, and the demand for luxury jewelry configuration will increase. When the economy is in a downturn, the demand for high-net-value households to allocate luxury jewelry will increase for the purpose of asset preservation and appreciation.

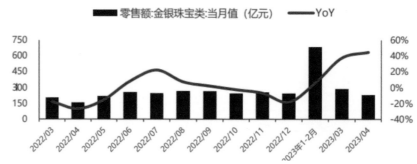

金银名贵珠宝类零售额
Gold and Silver Luxury Jewelry Retail Sales
资料来源：国家统计局，民生证券研究院
Source: National Bureau of Statistics, Minsheng Securities Research Institute

（二）名贵珠宝的市场动态
Market Dynamics of Luxury Jewelry

1.名贵珠宝投资要抓住市场发展趋势
Luxury Jewelry Investment Should Seize the Market Development Trend

当前，美国加息、俄乌冲突、巴以冲突导致国际形势复杂，国际金价走势强劲。这意味着，投资可以选择名贵珠宝类产品，以获得稳定的保值增值预期。首先，名贵珠宝的价值通常是极高的。名贵珠宝稀有且珍贵，因此被视为财富和社会地位的象征。其次，名贵珠宝也是一种避险工具。在经济不稳定时期，许多投资品的价格波动较大，而名贵珠宝的价格往往保持稳定。这使得投资者能够保持其资产价值，并在市场动荡时保持冷静。再次，名贵珠宝的流通性极强。这是投资者所追求的一个重要特性。无论是经济繁荣还是衰退时期，名贵珠宝都可以迅速变现。这些名贵珠宝在国际市场上容易找到买家，因此投资者可以随时将名贵珠宝转换为现金。

At present, the US interest rate hike, the Russia-Ukraine conflict, and the Palestinian-Israeli conflict have led to a complex international situation, and the international gold price has a strong increasing trend. This means that the investment can choose luxury jewelry products to obtain stable preservation and appreciation expectations. First, the value of luxury jewelry is usually extremely high. Luxury jewelry is rare and valuable, so it is regarded as a symbol of wealth and social status. Second, luxury jewelry is also a hedging tool. In times of economic uncertainty, the price of many investment goods is volatile, while the price of luxury jewelry tends to remain stable. This allows investors to preserve the value of their assets and remain calm in times of market turmoil. Third, luxury jewelry is highly liquid. This is an important characteristic sought after by investors. In good times and bad, luxury jewelry can be quickly cashed in. These precious jewels are easy to find buyers in the international market, so investors can convert precious jewels into cash at any time.

2.名贵珠宝投资要抓住名贵这个特征
Luxury Jewelry Investment Should Seize the Characteristics of Luxury

当前，中国国内消费结构和消费场景的变化，推动行业集中度、新零售模式和产品IP的创新发展。这意味着，首先，名贵珠宝投资要选择行业集中度更好的品牌，可以选择国潮文化IP类名贵珠宝，这样能够保障资产配置获得更高收益率。其次，名贵珠宝的美学价值不可忽视，它不仅仅是投资工具，更是艺术品和美的象征，能够增加生活品质。再次，名贵珠宝的保存性极佳。经过专业的切割和抛光，这些名贵珠宝变得坚硬且耐久，这意味着无论时间如何流转，名贵珠宝的价值不会因为时间的侵蚀而受损。与其它易损坏或易变质的投资品相比，名贵珠宝显然更具有长期保存的价值。最后，名贵珠宝可以作为遗产传承下去。随着家族一代代的繁衍，这些名贵珠宝成为了家族历史的见证和传承的象征，它们不仅仅是财富的象征，更是家族文化和历史的延续。

At present, the changes in China's domestic consumption structure and consumption scenarios have promoted the innovative development of industry concentration, new retail models and product IP. This means that, first of all, luxury jewelry investment should choose brands with better industry concentration, and you can choose luxury jewelry of traditional-culture IP related, which can guarantee higher returns for asset allocation. Secondly, the aesthetic value of luxury jewelry can not be ignored, it is not only an investment tool, but also a symbol of art and beauty, which can increase the quality of life. Thirdly, the preservation of luxury jewelry is excellent. Due to professional cutting and polishing, these luxury jewelry become hard and durable, which means that no matter how the passage of time, the value of luxury jewelry will not be damaged by the erosion of time. Compared with other easily damaged or easily spoiled investment goods, luxury jewelry obviously has more long-term preservation value. Finally, luxury jewelry can be passed on as a legacy. With the family's reproduction from generation to generation, these luxury jewelry have become the witness of family history and the symbol of inheritance. They are not only a symbol of wealth, but also the continuation of family culture and history.

（三）名贵珠宝的投资时机
Investment Opportunities for Luxury Jewelry

名贵珠宝的投资时机非常重要。在什么时候选择名贵珠宝投资，主要考虑投资保值因素、消费场景和产品工艺等因素。

The timing of luxury jewelry investment is very important. When to choose luxury jewelry investment, the main consideration is investment preservation factors, consumption scenarios and product processes.

1.名贵珠宝投资保值增值时机的选择
Choose the Opportunity for Luxury Jewelry Investment to Preserve and Increase Value

　　名贵珠宝投资，要关注国际金价的波动，金价波动会影响到名贵珠宝的价格。要关注人民币汇率波动的影响，在汇率波动较大的时候，选择名贵珠宝有利于资产保值。要关注其他投资资产的回报率，尤其是在房地产投资、股票投资、债市投资和汇市投资收益波动较大，甚至投资收益明显下跌的时候，投资名贵珠宝可以对冲其他资产价格下跌的风险。要关注投资消费场景，一般而言，仅仅满足婚恋消费需求，并不能有效实现名贵珠宝投资的增值。满足悦己需求和资产增值需求，可以通过投资名贵珠宝获得较好收益。要关注名贵珠宝工艺更新的时机。当前，名贵珠宝新工艺迭代升级，选择更具创新性的名贵珠宝产品，尤其是富含文化元素、文化附加值高、个性化和人文性特征更为明显的，这样更能获得较高投资收益。

　　Luxury jewelry investment, to pay attention to the fluctuations of the international gold price, as gold price fluctuations will affect the price of luxury jewelry. We need to pay attention to the impact of RMB exchange rate fluctuations, when the exchange rate fluctuations are large, the choice of luxury jewelry is conducive to asset preservation. We also need to pay attention to the rate of return of other investment assets, especially in real estate investment, stock investment, bond investment and foreign exchange investment income volatility, and even when investment returns significantly decline, investing in luxury jewelry can hedge the risk of falling prices of other assets. Do also pay attention to the investment and consumption scenario. In general, only to meet the needs of marriage and love consumption can not effectively achieve the appreciation of luxury jewelry investment. To meet your self-satisfying needs and asset appreciation needs, you can get better returns by investing in luxury jewelry. Pay attention to the timing of the renewal of the luxury jewelry process. At present, the new process of luxury jewelry is iterated and upgraded, and more innovative luxury jewelry products are selected, especially those rich in cultural elements, high cultural added value, and more obvious personalized and humanistic characteristics, so that higher investment returns can be obtained.

2.名贵珠宝投资要注重跨时套利策略
Luxury Jewelry Investment Should Pay Attention To Trans-time Arbitrage Strategy

　　所谓跨时套利，也就是预测在未来一个时期，某种物品会出现稀缺的情况，从而在低价买进，在未来高价卖出。名贵珠宝投资也是这个逻辑，这非常考验投资者对于未来市场趋势的判断能力，否则不会赚钱甚至会巨亏。《货殖列传》中叙述了范蠡用跨时套利的策略，成为当时首富的故事。

　　The so-called trans-time arbitrage is to predict that in the future period, a certain item will be scarce, so as to buy at a low price and sell at a high price in the future. Luxury jewelry investment is also the same logic, which is a very harsh test of investors' ability to judge the future market trend, otherwise they will not make money or even suffer huge losses. The story of Fan Li who became the richest man at that time by using the strategy of trans-time arbitrage is narrated in the Biography of Goods Colonization.

　　是否选择名贵珠宝作为跨时套利策略，受到宏观经济、社会环境的深刻影响。例如，在楚汉争霸时期，当所有人都选择储存名贵珠宝时，督道仓的守吏任氏选择存储粮食，最终因粮食短缺，任氏以粮食换名贵珠宝最终获得大量财富。这是当时战时极端情况下跨时套利的特殊例子，非常考验投资者对于未来的判断能力。然而，在当前总体和平的环境下，投资粮食并非优选项，社会剩余资金会选择投向名贵珠宝，作为未来财产保值增值的重要策略。

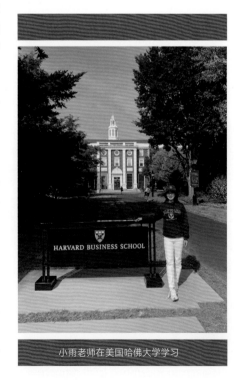

小雨老师在美国哈佛大学学习

Whether to choose luxury jewelry as a cross-time arbitrage strategy is deeply affected by macroeconomic and social environment. For example, during the period of Chu and Han Dynasties, when everyone chose to store luxury jewelry, Mr. Ren, the officer of Dodaocang, chose to store grain. In the end, due to the shortage of grain, Ren exchanged grain for luxury jewelry and eventually gained a lot of wealth. This is a special example of cross-time arbitrage under the extreme circumstances of the war at that time, and it is a great test of investors' ability to judge the future. However, in the current generally peaceful environment, investment in food is not the preferred option, and the remaining funds of the society will choose to invest in luxury jewelry as an important strategy for maintaining and increasing the value of property in the future.

四、影响名贵珠宝价值的因素
Factors that Influence the Value of Luxury Jewelry

（一）影响名贵珠宝价值的品质因素
Quality Factors Affecting the Value of Luxury Jewelry

1.颜色。名贵珠宝的颜色是一个关键的价值因素。鲜艳、纯净、均匀的颜色能够赋予珠宝独特的魅力，使其在众多珠宝中脱颖而出。例如，红宝石的鲜艳红色、蓝宝石的深邃蓝色以及祖母绿的柔和绿色，都是评判其价值的重要标准。然而，有时也会有特定颜色的名贵珠宝，更为珍贵。

Color. The color of luxury jewelry is a key value factor. Bright, pure and even colors can give jewelry a unique charm, making it stand out among many jewelries. For example, the vivid red of rubies, the deep blue of sapphires and the soft green of emeralds are important criteria for judging their value. However, sometimes there are luxury jewels in specific colors that are more precious.

2.净度。透明度和纯净度也是衡量名贵珠宝品质的重要依据。高品质的名贵珠宝需要具备高度透明且内部无瑕的特点。透明度高的名贵珠宝能够让光线充分穿透，呈现出璀璨的光泽。纯净度高的名贵珠宝则意味着其内部杂质极少，充分展现出珠宝的内在美。通常来说，越透明、越清晰的名贵珠宝价值越高。

Clarity. Transparency and purity are also important bases for measuring the quality of luxury jewelry. High quality luxury jewelry needs to be highly transparent and internally flawless. High transparency of the luxury jewelry can let the light fully penetrate, showing a bright luster. The high purity of luxury jewelry means that its internal impurities are very few, fully showing the inner beauty of the jewelry. Generally speaking, the more transparent and clear the jewelry, the higher the value.

3.切割和形状。珠宝的切割质量会影响其光学性能，包括反射和折射。切割的评价主要包括对称性、抛光程度和设计风格等方面。对称性是指珠宝各部分之间的比例关系，比例协调的珠宝更具完美外观。抛光程度是指珠宝表面的光滑程度，抛光好的珠宝表面更能呈现光泽。设计风格则是指珠宝的整体外观和美感。

Cut and shape. The quality of the cut of the jewelry affects its optical properties, including reflection and refraction. The evaluation of cutting mainly includes symmetry, polishing degree and design style. Symmetry refers to the proportional relationship between the various parts of the jewelry, the proportion of coordinated jewelry are having more perfect appearance. The degree of polishing refers to the smoothness of the jewelry surface, and the polished jewelry surface is more capable of showing luster. Design style refers to the overall appearance and beauty of the jewelry.

4.重量。名贵珠宝的重量通常以克拉为计量单位。一般来说，同等条件下，重量较大的名贵珠宝更为稀有，因此也更有价值。

Weight. The weight of luxury jewelry is usually measured in carats. In general, under the same conditions, the larger weight of the luxury jewelry, it will be rarer, and therefore more valuable.

5.产地。名贵珠宝的产地也会影响其价值。一些产自特定地区的名贵珠宝可能因为其独特的品质而更受欢迎。由于不同品质的名贵珠宝在自然界中的产地不同，产量各不相同，因此越稀有的名贵珠宝往往价值越高。例如，红宝石、蓝宝石、祖母绿等珍贵品种，因为产量稀少，价格一直居高不下。

Origin. The origin of a luxury jewel can also affect its value. Some luxury jewelry made in a specific region may be more popular because of its unique quality. Because of the different quality of luxury jewelry in nature in different places, the output is different, so the rarer the luxury jewelry is often the higher the value. For example, precious varieties such as rubies, sapphires and emeralds, as the production is scarce, the price has been always very high.

6.处理和改良。某些名贵珠宝可能经过热处理、辐射或其他方法进行改良。这些处理可能会影响名贵珠宝的价值。

Treatment and improvement. Some valuable jewelry may be modified by heat treatment, radiation or other methods. These treatments can affect the value of valuable jewelry.

7.品质证书。具有权威的名贵珠宝鉴定证书的名贵珠宝通常更容易销售，并且更受信任。

Certificate of quality. Expensive jewelry with an authoritative certificate of identification is usually easier to sell and more trusted.

（二）影响名贵珠宝价值的经济环境因素
Economic and Environmental Factors Affecting the Value of Luxury Jewelry

名贵宝石投资是否具有持续价值，与经济放缓、经济收缩息息相关。在房价下跌、企业利润下滑、资金外逃的情况下，经济增速会放缓，甚至收缩，这会导致大量社会剩余资金从产业部门挤出。在房市退热情况下，局部房市资金进入股市并不能支撑股市走强。此时，选择名贵宝石作为资产配置方式，能够为剩余资金找到安全的投资渠道，确保资产保值增值。

Whether the luxury jewelry investment has sustained value is closely related to the economic slowdown and economic contraction. In the case of falling housing prices, declining corporate profits, and capital flight, economic growth will slow down or even contract, which will lead to the effect of squeezing out a large amount of surplus social funds from the industrial sector. In the case of housing market fever, partial housing market funds into the stock market can not support the strength of the stock market. At this time, choosing luxury jewelry as an asset allocation method can find a safe investment channel for the remaining funds and ensure the preservation and appreciation of assets.

（三）综合评价名贵珠宝的价值
Comprehensive Evaluation of the Value of Luxury Jewelry

名贵珠宝价值的评价，要综合考虑市场因素、成本因素，以及收益因素。

The evaluation of the value of luxury jewelry should comprehensively consider market factors, cost factors and income factors.

1.影响名贵珠宝价值的市场因素
Market Factors Affecting the Value of Luxury Jewelry

名贵珠宝要获得更好的价值，一般应具有公开活跃的交易市场，可以通过足够的交易数据，来衡量不同级别名贵珠宝市场数据的可靠性。在衡量某种名贵珠宝价值的时候，可以搜集评估该类型名贵珠宝以往的市场交易信息，作为重要的参考标准。选择相似类型名贵珠宝作为参照物，可以提升名贵珠宝价值的可比性。选择合适的名贵珠宝交易市场作为评估依据，对于名贵珠宝的价值评估也非常重要。如果选定了目标珠宝，在评定该珠宝价值的时候，可以考虑它与参照物珠宝之间的区别，以及在名贵珠宝市场级别、名贵珠宝市场交易条件等因素上的差异，并分析整理相同或者相似名贵珠宝交易的信息资料，从而获得较为准确的定价。

小雨老师参加斯里兰卡宝石展

In order to obtain better value for luxury jewelry, there should generally be an open and active trading market, and the reliability of the market data of different levels of luxury jewelry can be measured through sufficient trading data. When measuring the value of a certain luxury jewelry, it is possible to collect and evaluate the past market transaction information of this type of luxury jewelry as an important reference standard. Choosing similar types of luxury jewelry as a reference can enhance the comparability of luxury jewelry values. Choosing the right luxury jewelry trading market as the basis for the evaluation of luxury jewelry value is also very important. If the target jewelry is selected, when assessing the value of the jewelry, it can be considered the difference between it and the reference jewelry, as well as the difference in the market level of luxury jewelry, luxury jewelry market trading conditions and other factors. Through analyzing and sort out the same or similar luxury jewelry trading information, a more accurate pricing can be obtained.

2.影响名贵珠宝价值的成本因素
Cost Factors Affecting the Value of Luxury Jewelry

在衡量名贵珠宝价值的时候，还要考虑名贵珠宝自身的成本。要分析名贵珠宝是否可以被再生产，名贵珠宝首饰是否可以被复制，也就是名贵珠宝是否具有稀缺性。要衡量名贵珠宝成本的核心构成要素，具体包括材料成本、加工制作成本、相关税费、合理利润等内容。要衡量名贵珠宝是否会随着时间的推移贬值，包括材质变化导致的实体性贬值、经济金融环境变化导致的经济性贬值，以及消费和投资需求变化导致的功能性贬值。

When measuring the value of luxury jewelry, it is also necessary to consider the cost of luxury jewelry itself. To analyze whether the luxury jewelry can be reproduced, whether the luxury jewelry can be copied, that is, which means whether or not the luxury jewelry has scarcity. To measure the core components of the cost of luxury jewelry, we need to take these into consideration: material costs, processing costs, related taxes, reasonable profits and other content. To measure whether luxury jewelry will depreciate over time, we need to consider substantive depreciation caused by material changes, economic depreciation caused by changes in the economic and financial environment, and functional depreciation caused by changes in consumption and investment demand.

Luxury Jewelry
Investment and Appreciation

第五章
05 Chapter Five
名贵珠宝
投资策略与风险管理
Luxury Jewelry Investment
Strategy and
Risk Management

一、名贵珠宝投资的短期与长期策略
Short-term and Long-term Strategies for Luxury Jewelry Investment

名贵珠宝投资要重点关注短期策略和长期策略，从而有效提升名贵珠宝投资收益率。在短周期内投资名贵珠宝，要看名贵珠宝投资市场是否景气，关注名贵珠宝的变现功能。在长周期内投资名贵珠宝，要看未来名贵珠宝市场增长空间是否持续显著，关注名贵珠宝长期投资的增值价值。

Luxury jewelry investment should focus on short-term strategies and long-term strategies, so as to effectively improve the return on investment of luxury jewelry. In a short period of investment in luxury jewelry, it depends on whether the luxury jewelry investment market is booming, and also pay attention to the realization function of luxury jewelry. Investing in luxury jewelry in a long period depends on whether the growth space of the luxury jewelry market in the future continues to be significant, and pays attention to the value-added value of long-term investment in luxury jewelry.

（一）名贵珠宝投资短期变现快但是增值很难
Luxury Jewelry Investment in the Short-term: Fast Realization But Difficult to Add Value

从短周期来看，居民人均可支配收入的涨跌会影响对名贵珠宝的消费需求，这种名贵珠宝消费需求具有极强的可选择性。在短周期内，居民可以购买名贵珠宝用于悦己、送礼、婚恋等消费需要。例如，2020年至2023年的新冠疫情期间，人们对名贵珠宝的消费需要有所下降，但是投资需求反而增加。在短周期内，名贵珠宝具有好交易、变现快的特点，但是短期增值的可能性比较低。

From a short cycle point of view, the rise and fall of per capita disposable income of residents will affect the consumption demand for luxury jewelry, which is highly selective. In a short period of time, residents can buy luxury jewelry for their own enjoyment, gifts, marriage and other consumption needs. For example, during the COVID-19 pandemic from 2020 to 2023, people's consumption of luxury jewelry had decreased, but investment demand had increased. In a short period of time, luxury jewelry has the characteristics of easy trading and fast realization, but the possibility of short-term appreciation is relatively low.

（二）名贵珠宝投资长期保值增值
Luxury Jewelry Investment in Long-term: Preservation and Appreciation

从长周期来看，居民人均可支配收入增长，会提升居民对于名贵珠宝的投资需求，以获得更高的投资收益。在长周期内，名贵珠宝具有比较稳定的价格上涨预期，居民可选择名贵珠宝作为资产配置方式。在经济企稳的时候，消费者可以增加对名贵珠宝的配置，提升资产配置的收益率。此外，如果遇到经济不稳、局势动荡等各种因素导致的资产价格短期迅速下跌的情况，消费者常常可以"抄底"名贵珠宝，在迅速扩大的差价中提高回报率。

In the long term, the growth of residents' per capita disposable income will increase residents' investment demand for luxury jewelry to obtain higher investment returns. In the long cycle, luxury jewelry has a relatively stable price rise expectation, and residents can choose luxury jewelry as an asset allocation method. When the economy stabilizes, consumers can increase the allocation of luxury jewelry and improve the return on asset allocation. In addition, in the event of a rapid short-term decline in asset prices caused by various factors such as economic instability and instability, consumers can often "buy at lowest price" of luxury jewelry and increase returns in the rapidly expanding price difference.

欧泊

Luxury Jewelry Investment and Appreciation

Luxury jewelry investment should focus on short-term strategies and long-term strategies, so as to effectively improve the return on investment of luxury jewelry.

（三）名贵珠宝投资的短期策略
Short-term Strategies for Luxury Jewelry Investment

在消费端，消费者可以考虑配置名贵珠宝，有效改善在婚恋、悦己、投资等方面的名贵珠宝产品结构。在投资端，消费者在短周期内购买名贵珠宝，应重点关注名贵珠宝的抗风险功能和抗通货膨胀属性。

当前，面对俄乌冲突和巴以冲突等地缘风险，名贵珠宝的投资需求迅速上升，以满足资产避险需求。新冠疫情后，全球各国央行大多数采取了宽松的货币政策，全球通货膨胀比较严重。在这种情况下，消费者可以考虑配置一定比例的名贵珠宝，作为抗通胀的资产配置方案。

At the consumer end, consumers can consider the allocation of luxury jewelry, effectively improve the structure of luxury jewelry products in marriage, love, happiness, investment and other aspects. On the investment side, consumers buying luxury jewelry in a short period of time should focus on the anti-risk function and anti-inflation properties of luxury jewelry.

At present, in the face of geopolitical risks such as the Russia-Ukraine conflict and the Israeli-Palestinian conflict, the investment demand for luxury jewelry has risen rapidly to meet the demand for asset hedging. After the COVID-19 epidemic, most central banks around the world have adopted loose monetary policies, and global inflation is relatively serious. In this case, consumers may consider allocating a certain percentage of luxury jewelry as an inflation-proof asset allocation scheme.

（四）名贵珠宝投资的长期策略
Long-term Strategies for Luxury Jewelry Investment

能否在长周期配置名贵珠宝，关键要看名贵珠宝市场在长期是否有稳健增长和增值的空间。就中国市场而言，中产群体扩大、消费结构升级，对名贵珠宝等高端消费的需求大幅增加。此外，国潮背景下的传统文化消费，使名贵珠宝成为不可或缺的优良载体，名贵珠宝获得更为强劲的增值空间。这意味着，中国名贵珠宝市场是一个长期稳定增长的行业，具有庞大的市场规模，以及较高的收益弹性空间。投资名贵珠宝，在长期可以获得较为丰厚的投资回报。

Whether the luxury jewelry can be configured in a long period depends on whether the luxury jewelry market has room for steady growth and appreciation in the long run. As far as the Chinese market is concerned, the expansion of the middle class and the upgrading of the consumption structure have significantly increased the demand for high-end consumption such as luxury jewelry. In addition, the traditional cultural consumption under the background of traditional-culture tide has made luxury jewelry an indispensable and excellent carrier, and luxury jewelry has gained a stronger value-added space. This means that China's luxury jewelry market is a long-term stable growth of the industry, with a huge market size, as well as a high yield flexibility space. Investment in luxury jewelry, in the long run can obtain a relatively rich return on investment.

二、投资组合构建策略
Investment Portfolio Construction Strategy

（一）资产配置的目标和种类
Objectives and Types of Asset Allocation

投资名贵珠宝，要做好投资组合规划。一般而言，资产配置要求实现资产隔离目标、财富规划目标和财富传承目标。投资的核心是在管控资产风险的基础上，合理把握投资周期和资产结构，实现资产保值增值。

在资产配置的种类上，一般包括现金资产、资管计划、保险金型投资、房地产/不动产、股权直投或私募股权基金、债权及固定收益类产品、股票及权益类产品、黄金、艺术品、新兴数字资产等。从长远来看，保值增值仍然是资产配置的核心属性，名贵珠宝投资同样是未来资产配置的重要内容。

To investment in luxury jewelry, a good portfolio planning is essential. In general, asset allocation requires the achievement of asset segregation goals, wealth planning goals, and wealth inheritance goals. The core of investment is to reasonably grasp the investment cycle and asset structure on the basis of asset risk control, so as to maintain and increase the value of assets.

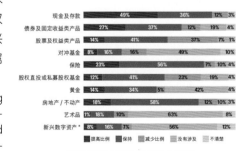

2022年和2021年资产组合调整
资料来源：财富研究院
Portfolio Adjustments for 2022 and 2021
Source: Wealth Research Institute

In terms of the types of asset allocation, it generally includes cash assets, asset management plans, insurance investments, real estate/real estate, direct equity or private equity funds, debt and fixed income products, stocks and equity products, gold, artwork, emerging digital assets, etc. In the long run, preserving and increasing value is still the core attribute of asset allocation, and luxury jewelry investment is also an important part of future asset allocation.

(二)以名贵珠宝投资实现多元化资产配置
Diversified Asset Allocation with Luxury Jewelry Investment

1.以名贵珠宝平衡投资策略
Balance Your Investment Strategy with Luxury Jewelry

当前，经济受挫、市场波动较大，投资理财的风险偏好降低。投资者的风险管理意识提升，更多居民提高了家庭备用金，提高了家庭财产保险和年金类保险配置比例，理财偏好比较保守、固收类理财增加。建议做好资产多元化配置，平衡好保本投资策略、均衡投资策略和增长投资策略之间的关系，在增长投资策略部分选配名贵珠宝资产，提高资产收益率。

At present, the economy is frustrated, the market is volatile, and the risk appetite of investment and financial management is reduced. Investors' awareness of risk management has increased, more residents have increased their household reserve funds, increased the proportion of household property insurance and annuity insurance allocation, and their financial preference is more conservative and fixed-income financial management has increased. It is suggested to do a good job in asset diversification, to balance the relationship between capital preservation investment strategy, balanced investment strategy and growth investment strategy, and select luxury jewelry assets in the growth investment strategy to improve the return on assets.

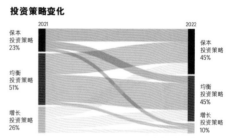

2021-2022年投资策略变化图
资料来源：财富研究院
2021-2022 Investment Strategy Changes Chart
Source: Wealth Research Institute

2.提高名贵珠宝配置比例、优化家庭财务目标
Increase the Proportion of Luxury Jewelry Allocation and Optimize Damily Financial Goals

建议将家庭资产划分为家庭备用金、保障配置和增值配置部分。

一是要配置好家庭备用金。这部分以现金、存款和货币基金为主，备用金一般为家庭月度开支的3-6倍，也就是要确保家庭3个月到半年左右的日常开支。但是，由于备用金会挤占资金利用效率、降低收益率，所以备用金并非越高越好。

二是要优化保障配置。这部分主要用于购置财产险和年金险，适度提高购置比例，满足个性化、多样化的商业保险需求。但是，保障配置部分过高，同样会降低资产利用效率和收益率。

三是做好增值配置。这部分主要用于改善长期投资收益，提高资金利用率和收益率。当前，经济下行，投资房地产、期权期货类金融衍生品、信托、私募基金、股票、权益类理财，其收益率明显下滑。建议在另类资产中提高名贵珠宝的配置比例，利用其稀有性、增值性特点，提升资产收益，降低资产管理风险。要培养长期投资理念及行为，了解资产供应变革趋势，通过名贵珠宝资产配置带来长期收益。

It is suggested to divide household assets into household reserve, security allocation and value-added allocation.

The first is to allocate a good family reserve fund. This part is mainly cash, deposits and monetary funds, and the reserve fund is generally 3-6 times the monthly expenses of the family, that is, to ensure the daily expenses of the family for about 3 months to half a year. However, as the reserve fund will crowd out the efficiency of the utilization of funds and reduce the rate of return, the higher the reserve fund does not result in better return.

Second, we need to optimize the security allocation. This part is mainly used for the purchase of property insurance and annuity insurance, and moderately increase the purchase ratio to meet the personalized and diversified commercial insurance needs. However, if the security allocation part is too high, it will also reduce the efficiency of asset utilization and the rate of return.

Third, we will make value-added arrangements. This part is mainly used to improve long-term investment returns, improve the utilization rate of funds and yield. At present, the economic downturn, investment in real estate, options and futures financial derivatives, trust, private equity funds, stocks, equity financing, its return has declined significantly. It is suggested to increase the allocation proportion of luxury jewelry in alternative assets, make use of its rarity and value-added characteristics, to improve asset returns and reduce asset management risks. It is necessary to cultivate long-term investment concepts and behaviors, understand the changing trend of asset supply, and bring long-term returns through luxury jewelry asset allocation.

居民投资理财配置最多的三种资产
资料来源：《中国居民投资理财行为调研报告》
The Three Most Common Assets Allocated by Residents for Investment and Financial Management
Source: "Research Report on Investment and Financial Management Behavior of Chinese Residents"

3.提高名贵珠宝的配置比例的原因
Reasons for Increasing the Proportion of Luxury Jewelry

从投资收益看，名贵珠宝更具较高的投资盈利能力。短期内配置黄金及其饰品，能够在金价走高的时期获得较好的收益。从长期看，黄金及其饰品仅具备保值功能，在溢价能力方面相对较弱，这就要求在配置名贵珠宝方面下功夫。因此，可以选择溢价能力更高的名贵珠宝作为资产配置。值得注意的是，要选择具有高端品牌地位、产品研发能力强、新零售渠道强的名贵珠宝，从而获得更高的溢价收益。

From the perspective of investment income, luxury jewelry has higher investment profitability. Short-term allocation of gold and its accessories can obtain better returns in the period of high gold prices. In the long run, gold and its jewelry only have the function of preserving value, and are relatively weak in terms of premium ability, which requires efforts in the configuration of luxury jewelry. Therefore, luxury jewelry with higher premium ability can be selected as an asset allocation. It is worth noting that it is necessary to choose luxury jewelry with high-end brand status, strong product research and development ability, and good new retail channels, so as to obtain higher premium income.

（三）名贵珠宝投资需要考虑的具体因素
Specific Factors that Need to be Considered for Luxury Jewelry Investment

构建名贵珠宝投资组合的策略涉及多个因素，包括名贵珠宝的类型、品质、市场趋势、风险管理等。

第一，了解名贵珠宝市场。在构建投资组合之前，深入了解名贵珠宝市场的基本知识至关重要。了解不同类型的名贵珠宝，了解名贵珠宝的产地、品质评估标准，以及市场价格波动的因素。

第二，选择多样化投资。类似于金融投资组合，名贵珠宝投资组合也应该实现多样化。不要将所有资金投入到同一种类的名贵珠宝中。要考虑配置包含不同类型和颜色的名贵珠宝，以分散风险。

第三，考虑名贵珠宝的投资级别。名贵珠宝可以根据其投资级别进行分类，例如，投资级别的钻石、红蓝宝石等。高品质、稀有和大型的名贵珠宝通常更受投资者青睐，但也伴随着更高的价格和风险。

红宝石耳钉

第四，关注品质和证书。选择高品质、有证书的名贵珠宝是构建投资组合的关键。Gemological Institute of America(GIA)等机构提供的证书可以确保名贵珠宝的品质和真实性。

第五，分析市场趋势。跟踪名贵珠宝市场的趋势和预测未来的发展对投资决策至关重要。了解名贵珠宝的供需关系、产地变化和流行趋势。

第六，定期评估和更新。名贵珠宝市场和经济状况都可能发生变化，因此定期评估投资组合的表现并进行必要的调整是重要的。有时可能需要出售一些名贵珠宝以获得利润或调整投资组合的风险水平。

第七，充分考虑流动性。名贵珠宝相对于其他投资工具(如股票、债券)可能缺乏流动性。在构建名贵珠宝投资组合时，要考虑到名贵珠宝的买卖难度以及市场变动对其流动性的影响。

第八，把握风险管理。对于任何投资组合，风险管理都是关键。了解自己的风险承受能力，避免过度集中在某一种名贵珠宝类型或市场上。

第九，寻求专业建议。由于名贵珠宝市场相对较小且具有特殊性，寻求专业的名贵珠宝评估师或顾问的建议是明智之举。他们可以提供有关市场趋势、品质评估和投资策略的有价值的意见。

第十，关注可持续性。考虑选择符合可持续发展标准的名贵珠宝，以适应当代社会对可持续性和道德采购的关注。

值得注意的是，名贵珠宝投资属于相对较为特殊和非流通性较高的投资类型，因此需要更谨慎的考虑和研究。

The strategy for building a luxury jewelry portfolio involves a number of factors, including the type of luxury jewelry, quality, market trends, risk management, etc.

First, understand the luxury jewelry market. Before building a portfolio, it is essential to have a deep understanding of the basics of the luxury jewelry market. Learn about the different types of luxury jewelry, learn about the origin of luxury jewelry, quality evaluation criteria, and market price fluctuations.

Second, diversify your investments. It is similar to a financial portfolio, a luxury jewelry portfolio should also be diversified. Don't put all your money into the same kind of jewelry. Consider configuring valuable jewelry with different types and colors to spread the risk.

Third, consider the investment grade of luxury jewelry. Luxury jewelry can be classified according to its investment grade, for example, investment grade diamonds, red sapphires, etc. High quality, rare and large jewelry is generally preferred by investors, but also comes with higher prices and risks.

小雨老师拜访斯里兰卡最大的宝石矿主

Fourth, focus on quality and credentials. Choosing high-quality, certified jewelry is key to building a portfolio. Certificates from institutions such as the Gemological Institute of America (GIA) can ensure the quality and authenticity of valuable jewelry.

Fifth, analyze market trends. Tracking trends in the luxury jewelry market and predicting future developments are crucial to investment decisions. Understand the supply and demand relationship, origin changes and fashion trends of luxury jewelry.

Sixth, evaluate and update regularly. Both the luxury jewelry market and economic conditions are subject to change, so it is important to regularly evaluate the performance of your portfolio and make the necessary adjustments. Sometimes it may be necessary to sell some valuable jewelry for profit or to adjust the risk level of the portfolio.

Seventh, fully consider liquidity. Luxury jewelry may be illiquid relative to other investment instruments (such as stocks, bonds). When constructing a luxury jewelry portfolio, it is necessary to consider the difficulty of buying and selling luxury jewelry and the impact of market changes on its liquidity.

Eighth, grasp risk management. Risk management is key to any portfolio. Know your risk tolerance and avoid being overly concentrated in one type of luxury jewelry or market.

Ninth, seek professional advice. Due to the relatively small and specific nature of the luxury jewelry market, it is wise to seek the advice of a professional luxury jewelry appraiser or consultant. They can provide valuable advice on market trends, quality assessments and investment strategies.

Tenth, focus on sustainability. Consider choosing luxury jewelry that meets sustainability standards to meet contemporary concerns about sustainability and ethical sourcing.

It is worth noting that luxury jewelry investment is a relatively special and non- high liquidity investment type, so it needs more careful consideration and research.

（四）名贵珠宝投资/投机风险评级表
Luxury Jewelry Investment/Speculative Risk Rating Table

表1：名贵珠宝投资/投机风险评级表
Table 1: Luxury Jewelry Investment/Speculative Risk Rating Table

名贵宝石风险等级 \ 投资者类型	保守型(C1)	谨慎型(C2)	稳健型(C3)	积极型(C4)	激进型(C5)
低风险（R1）	√	√	√	√	√
中低风险（R2）	×	√	√	√	√
中等风险（R3）	×	×	√	√	√
中高风险（R4）	×	×	×	√	√
高风险（R5）	×	×	×	×	√

第一，净资产3000万元人民币以下家庭的名贵珠宝投资类型和风险类型。

总体上看，净资产3000万元人民币以下的家庭，更关注家庭日常消费等常规项目，风险承受能力相对较低，更关注现金流，保值能力、变现能力、现金偿付能力和抗风险能力。具体而言，净资产600万元人民币以下的中产家庭，属于保守型(C1)，名贵珠宝投资偏向低风险(R1)，建议关注保值能力和变现能力。净资产600万元人民币至1000万元人民币的富裕家庭，属于谨慎型(C2)，名贵珠宝投资偏向中低风险(R2)，建议关注保值能力、变现能力和现金偿付能力。净资产1000万元人民币至3000万元人民币的较高净值家庭，属于稳健型(C3)，名贵珠宝投资偏向中等风险(R3)，建议关注保值能力、变现能力、现金偿付能力和抗风险能力。

第二，净资产3000万元人民币元以上家庭的名贵珠宝投机类型和风险类型。

总体上看，净资产3000万元人民币以上的高净值家庭，现金流并非首要考虑因素，更加关注高品质生活、奢侈消费和家族财富传承，风险承受能力较高，关注投资能力和传承能力。具体而言，净资产3000万元人民币至2亿元人民币的超高净值家庭，属于积极型(C4)，名贵珠宝投机偏向中高风险(R4)，建议关注保值能力、抗风险能力和投资能力。净资产2亿元人民币以上（约3000万美元以上）的国际超高净值家庭，属于激进型(C5)，名贵珠宝投机偏向高风险(R5)，建议关注保值能力、抗风险能力、投资能力和传承能力。

First, the investment types and risk types of luxury jewelry for families with net assets below 30 million yuan.

Generally speaking, households with net assets of less than 30 million yuan pay more attention to routine items such as household daily consumption, and their risk tolerance is relatively low. They pay more attention to cash flow, and their preservation ability, liquidity ability, cash solvency and anti-risk ability are more important. Specifically, middle-class families with net assets of less than 6 million yuan belong to the conservative type (C1), and the investment of luxury jewelry tends to be low-risk (R1), so it is recommended to pay attention to the preservation ability and liquidity ability. Wealthy families with net assets of 6 million yuan to 10 million yuan belong to the prudent type (C2), and the investment of luxury jewelry is inclined to medium and low risk (R2), and it is recommended to pay attention to the ability to maintain value, liquidity and cash solvency. Higher net value families with net assets of 10 million yuan to 30 million yuan belong to the stable type (C3), and luxury jewelry investment is inclined to medium risk (R3), and it is recommended to pay attention to the ability to maintain value, liquidity, cash solvency and anti-risk ability.

Second, the types of speculation and risk of luxury jewelry in families with net assets of more than 30 million yuan.

In general, high net value families with net assets of more than 30 million yuan, cash flow is not the primary consideration. They are paying more attention to high-quality life, luxury consumption and family wealth inheritance, high risk tolerance, and more attention to investment ability and inheritance ability. Specifically, ultra-high net value families with net assets of 30 million yuan to 200 million yuan belong to the active type (C4), and the speculation bias of luxury jewelry is medium-high risk (R4), and it is recommended to pay attention to the ability to preserve value, resist risk and investment ability. International ultra-high net value families with net assets of more than 200 million yuan (about $30 million or more) belong to the aggressive type (C5), and luxury jewelry speculation tends to be high-risk (R5). It is recommended to pay attention to the ability to preserve value, resist risk, investment ability and inheritance ability.

三、名贵珠宝投资的风险评估
Risk Assessment of Luxury Jewelry Investment

（一）名贵珠宝投资与股市、债市和房地产投资的风险比较
Risk Comparison Between Luxury Jewelry Investment and Stock Market, Bond Market and Real Estate Investment

总体而言，债市、股市和货币政策、房地产等经济环境息息相关，是一体两面的，其投资的风险相对较高。当前，疫情的疤痕效应对经济的长期拖累，投资者在资产管理方面，化债优于支出，防风险重于稳增长，资产安全重于资产增值，固定资产投资增速放缓，居民投资远没有恢复，民间投资意愿不强。此外，中国货币政策通过降息扩大投放，释放流动性，这引起债券市场利率下降、债券价格上涨。但是，股票市场涨幅却不明显，因为股市除了较大的资金流入，股市还需要人们对未来经济利润增长的预期和信心，股市也深受汇率波动引起的海外资本观望的影响。资产配置要面对市场上有效需求不足的问题。在经济结构转型期，投资者要更新和重塑经济增长的知识结构，也要更新资产配置的知识结构。

在这种背景下，居民对于股市、债市等金融市场的合理预期比较低，对房地产市场的合理预期也比较低。这要求在金融市场波动比较大的时期，在各类资产配置总体增长转弱的时期，关注资产再配置的长期结构性变化，找到有效规避风险的合理资产。这种有效规避风险的资产就是名贵珠宝。做资产投资，要始终关注经济基本面、行业基本面和企业基本面。在经济基本面比较差、比较疲软的同时，一些行业基本面和企业基本面也可能表现较好。例如，在疫情防控期间缺乏其他有效资产投向的时候，名贵珠宝就成了资产保值增值的不二选择，这集中体现在疫情防控期间名贵珠宝的销量增速上。资产投资的关键，是看清经济周期的本质，找准不同时期的精准投向。

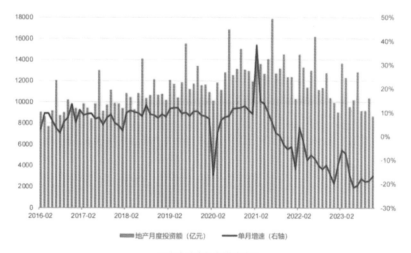

国内房地产投资增速降低
资料来源：Wind，摩根士丹利基金2024年度投资策略会
Domestic Real Estate Investment Growth Rate Decreases
Source: Wind, Morgan Stanley Fund 2024 Annual Investment Strategy Conference

In general, the bond market, the stock market and monetary policy, real estate and other economic environment are closely related, is one of two sides, its investment risk is relatively high. At present, the scar effect of the epidemic on the long-term drag on the economy, for investors in asset management, debt is better than spending, risk prevention is more important than stable growth, asset safety is more important than asset appreciation, fixed asset investment growth slowed down, residential investment is far from recovery, private investment willingness is not strong. In addition, China's monetary policy expands supply by cutting interest rates to release liquidity, which causes interest rates in the bond market to fall and bond prices to rise. However, the rise of the stock market is not obvious, because in addition to large capital inflows, the stock market also needs people's expectations and confidence in future economic profit growth, and the stock market is also deeply affected by overseas capital wait-and-see caused by exchange rate fluctuations. Asset allocation must face the problem of insufficient effective demand in the market. In the transition period of economic structure, investors should update and reshape the knowledge structure of economic growth, and also update the knowledge structure of asset allocation.

（二）名贵珠宝投资的风险缓释功能
The Risk Mitigation Function of Luxury Jewelry Investment

就名贵珠宝投资的风险而言，一般包括产品迭代风险、库存减值风险、渠道变革风险。名贵珠宝投资在各种投资类型中是比较特殊的种类，能否有效规避这些风险，是决定名贵珠宝投资能否获得较高收益的关键因素。

As far as the risk of luxury jewelry investment is concerned, it can be generally included as: product iteration risk, inventory impairment risk and channel change risk. Luxury jewelry investment is a special type in various investment types. Whether it can effectively avoid these risks is the key factor to determine whether luxury jewelry investment can obtain higher returns.

1.名贵珠宝投资降低产品迭代风险
Luxury Jewelry Investment Reduces Product Iteration Risk

相较于其他投资产品类型，名贵珠宝投资能够有效抵御产品迭代风险的冲击，获得较好的长期稳定收益。在科学技术不断革新的条件下，工业品和消费品的迭代更新速度会非常快，例如手机、电脑、照相机、汽车等一般工业品和消费品。科学技术的迅速革新甚至会带来工业品、消费品的跨界竞争，智能手机挤占相机市场就是例子。在其他投资领域，需要时刻关注科技进步对产品迭代的影响，关注科技进步对产业格局和行业格局的影响。但是，名贵珠宝的投资，很少面临产品迭代风险。因为，人们对名贵珠宝的关注焦点主要是符号价值，而非实际使用功能，名贵珠宝的实际使用功能被迭代的风险相对较小，甚至几乎为零。即使是名贵珠宝领域的技术进步，也不会导致名贵珠宝行业整体迭代的颠覆性风险。

Compared with other investment product types, luxury jewelry investment can effectively resist the impact of product iteration risks and obtain better long-term stable returns. Under the condition of continuous innovation of science and technology, the iterative update speed of industrial products and consumer goods will be very fast, such as mobile phones, computers, cameras, cars and other general industrial products and consumer goods. The rapid innovation of science and technology will even bring cross-border competition between industrial and consumer goods, such as the encroachment of smart phones on the camera market. In other areas of investment, it is necessary to always pay attention to the impact of scientific and technological progress on product iteration, and pay attention to the impact of scientific and technological progress on the industrial pattern and industry pattern. However, the investment of luxury jewelry rarely faces the risk of product iteration. Because people's focus on luxury jewelry is mainly symbolic value, rather than the actual use function. The risk of the actual use function of luxury jewelry is relatively small, or even almost zero. Even the technological progress in the field of luxury jewelry will not lead to the disruptive risk of the overall iteration of the luxury jewelry industry.

2.名贵珠宝投资降低库存减值风险
Luxury Jewelry Investment Reduces Inventory Impairment Risk

名贵珠宝由于其自身的材质稳定性，易于长期库存且不变质，折旧率比较低，能够有效抵御库存减值风险。同时，名贵珠宝具有稀缺性，通常更容易和期货、期权、租赁业务等金融工具挂钩，可以通过金融工具对冲降低名贵珠宝原材料的价格起伏波动的影响，从而更好地降低库存减值风险。此外，名贵珠宝的折旧率非常低，这节约了回收再造的成本，非常利于保值增值。

Due to its own material stability, luxury jewelry is easy to long-term inventory and does not deteriorate, and the depreciation rate is relatively low, which can effectively resist the risk of inventory impairment. At the same time, luxury jewelry is scarce, and it is usually easier to link with financial instruments such as future goods, options and leasing business. Financial instruments can be used to hedge to reduce the impact of price fluctuations of luxury jewelry raw materials, so as to better reduce inventory impairment risk. In addition, the depreciation rate of luxury jewelry is very low, which saves the cost of recycling and is very conducive to maintaining and increasing value.

3. 名贵珠宝投资降低渠道变革风险

Luxury Jewelry Investment Reduces the Risk of Channel Change

一般而言，线上销售渠道主要针对不太要求体验感的产品，通过长尾客户获得更多利润，消除信息不对称和中间环节，使商品价格更低。但是，越是要求体验感的商品，受到线上渠道影响的可能性就越小。名贵珠宝就是属于强体验感的商品，更关注单颗名贵珠宝给购买者带来的情感体验、身份体验、社会地位体验和投资体验，很难通过线上销售的长尾效应压低价格。这意味着，名贵珠宝的线下销售渠道仍是主流方式，这更利于稳定名贵珠宝的品牌效应，也维持了名贵珠宝保值增值的价格稳定效应。

In general, online sales channels are mainly for products that do not require much experience, obtaining more profits through long-tail customers, eliminating information asymmetry and intermediate links, and making commodity prices lower. However, the more experiential a product is, the less likely it is to be affected by online channels. Luxury jewelry is a commodity with a strong sense of experience. It pays more attention to the emotional experience, identity experience, social status experience and investment experience brought by a single luxury jewelry to the buyer, and it is difficult to lower the price through the long tail effect of online sales. This means that the offline sales channel of luxury jewelry is still the mainstream way, which is more conducive to stabilizing the brand effect of luxury jewelry, and also maintains the price stability effect of maintaining and increasing the value of luxury jewelry.

4. 名贵宝石投资降低资产减值风险

Luxury Jewelry Investment to Reduce Asset Impairment Risk

名贵珠宝投资有利于高净值家庭提高风险意识、积极调整资金配置结构，尤其是在经济事项和法律事项中降低资产减值风险。从资产配置多元化的角度看，名贵珠宝投资能够防范风险、避免暴雷。例如，在婚姻中一方自己购买的名贵珠宝在法律上属于个人独立财产，不受离婚诉讼分割家庭共同财产的影响，不会导致个人资产减值。在家族办公室业务和家族信托业务中，现金流管理的主要目标是保障家族财富安全传承、保障现有生活质量、确保财富保值升值，名贵珠宝投资既能够保障家庭财产保值升值，又能够保障家族财产传承、精神传承和情感传承。

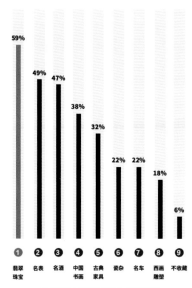

高净值人群最青睐的收藏品类
资料来源：2023 胡润至尚优品-中国高净值人群品牌倾向报告
Preferred Collectible Categories among High-Net-Worth Individuals
Source: 2023 Hurun-Tsinghua Luxury Brand Preferences Report for Chinese High-Net-Worth Individuals

Luxury jewelry investment is conducive to high net value families to improve risk awareness, actively adjust the capital allocation structure, especially in economic matters and legal matters to reduce the risk of asset impairment. From the perspective of asset allocation diversification, luxury jewelry investment can prevent risks and avoid thunderstorms. For example, expensive jewelry purchased by one party in a marriage is legally an independent property of the individual, and is not affected by the separation of the family's common property by divorce proceedings, and will not lead to the impairment of personal assets. In family office business and family trust business, the main goal of cash flow management is to ensure the safe inheritance of family wealth, guarantee the existing quality of life, and ensure the preservation and appreciation of wealth. Luxury jewelry investment can not only guarantee the preservation and appreciation of family property, but also guarantee the inheritance of family property, spiritual inheritance and emotional inheritance.

四、名贵珠宝投资的风险管理
Risk Management of Luxury Jewelry Investment

名贵珠宝投资虽然具有吸引力，但也伴随着一定的风险。在进行名贵珠宝投资时，有效的风险管理至关重要。

While investing in luxury jewelry is attractive, it also comes with certain risks. Effective risk management is essential when investing in luxury jewelry.

（一）关注市场价格波动风险
Pay Attention to the Risk of Market Price Fluctuations

名贵珠宝市场价格受供需、市场趋势、货币贬值和宏观经济因素等多方面影响，要避免这些因素导致的价格较大波动。在风险管理策略上，要定期跟踪名贵珠宝市场趋势，了解名贵珠宝价格的动向。多样化投资组合，不过度依赖某一种类型的名贵珠宝。要关注宏观经济趋势，及时调整投资策略。在不确定的经济环境下，谨慎选择投资时机。

The market price of luxury jewelry is affected by supply and demand, market trend, currency depreciation and macroeconomic factors, etc. It is necessary to avoid large price fluctuations caused by these factors. In terms of risk management strategy, it is necessary to regularly track the market trend of luxury jewelry and understand the price trend of luxury jewelry. Diversify your portfolio without over-reliance on one type of precious jewelry. It is necessary to pay attention to macroeconomic trends and adjust investment strategies in a timely manner. In an uncertain economic environment, choose your investment time carefully.

（二）关注流动性风险
Pay Attention to Liquidity Risks

名贵珠宝市场相对较小，要关注流动性，提升变现能力。在风险管理策略方面，要考虑投资较好流通的名贵珠宝，同时做好长期持有的准备。在需要变现时，关注名贵珠宝拍卖和交易动态。

The luxury jewelry market is relatively small, and it is necessary to pay attention to liquidity and improve liquidity. In terms of risk management strategy, it is necessary to consider investing in valuable jewelry that is better circulated and be prepared to hold it for a long time. When you need to cash out, pay attention to the auction and trading dynamics of luxury jewelry.

Investment Risk and Strategy

小雨老师与导师清华美院的唐旭祥教授交流

（三）关注伪造和品质优劣风险
Pay Attention to Counterfeiting and Quality Risks

名贵珠宝市场存在伪造风险，有可能购买到质量较差或伪造的名贵珠宝。在风险管理策略方面，要仔细选择可信赖的名贵珠宝商和交易平台，确保名贵珠宝具有相关的证书，定期进行名贵珠宝质量检验。

There is a risk of counterfeiting in the luxury jewelry market, and it is possible to buy poor quality or forged luxury jewelry. In terms of risk management strategy, it is necessary to carefully select trusted precious jewelers and trading platforms, ensure that luxury jewelry has relevant certificates, and regularly conduct quality inspection of luxury jewelry.

Investment Risk and Strategy

小雨在清华美院首饰专业学习

（四）关注保管和安全风险
Pay Attention to Storage and Security Risks

对于高价值的名贵珠宝，保管和安全是一个挑战，存在被盗窃和损坏的风险。在风险管理策略方面，要选择安全可靠的保管设施，购买足够的保险，确保名贵珠宝的安全。避免在家中储存大量高价值的名贵珠宝。

For high-value luxury jewelry, custody and security is a challenge, with risks of theft and damage. In terms of risk management strategy, it is necessary to choose safe and reliable storage facilities and purchase sufficient insurance to ensure the safety of luxury jewelry. And you should avoid storing large amounts of high-value jewelry in your home.

Investment Risk and Strategy

小雨在清华美院首饰专业学习

第六章
06 Chapter Six

名贵珠宝投资
顾问服务
Luxury Jewelry
Investment Advisory Services

一、名贵珠宝投资顾问服务中的核心内容
The Core Content of Luxury Jewelry Investment Advisory Services

（一）名贵珠宝投资顾问的概念
The Concept of Luxury Jewelry Investment Consultant

名贵珠宝投资顾问服务，是名贵珠宝投资顾问服务商向客户提供的专业化资产配置咨询服务，包括财务分析与规划、投资建议、名贵珠宝投资产品推介等专业化服务。在名贵珠宝投资顾问服务活动中，客户根据名贵珠宝投资服务商提供的投资顾问服务管理和运用资金，并降低投资风险，提高投资收益率。

Luxury jewelry investment advisory service is a professional asset allocation advisory service provided by luxury jewelry investment advisory service providers to customers, including financial analysis and planning, investment advice, luxury jewelry investment product promotion and other professional services. In the luxury jewelry investment advisory service activities, clients manage and use funds according to the investment advisory service provided by luxury jewelry investment service providers, reduce investment risks and improve investment returns.

（二）名贵珠宝投资顾问服务的目标和特点
Objectives and Characteristics of Luxury Jewelry Investment Advisory Services

名贵珠宝投资服务，是一个循序渐进的有机投资流程。对客户而言，其根本目的在于，通过财务目标分析、财务策略规划、资产投资建议、名贵珠宝投资建议服务，降低财务风险、降低对财务状况的焦虑，从而实现财务自由。对名贵珠宝服务商而言，可以通过名贵珠宝投资顾问服务实现客户关系管理目标，实现顾问式销售，从而提高名贵珠宝投资经营业绩。

Luxury jewelry investment service is a step-by-step organic investment process. For clients, its fundamental purpose is to reduce financial risks and anxiety about financial status through financial target analysis, financial strategy planning, asset investment advice, and luxury jewelry investment advice services, so as to achieve financial freedom. For luxury jewelry service providers, they can achieve customer relationship management objectives and advisory sales through luxury jewelry investment advisory services, so as to improve the investment and business performance of luxury jewelry.

名贵珠宝投资服务具有顾问特征，仅向客户提供名贵珠宝投资专业化建议，而不涉及客户资产配置的具体操作，客户拥有财务资源的最终决策权。名贵珠宝投资顾问服务具有专业特征，一是从业人员要有扎实的名贵珠宝专业知识，对名贵珠宝的种类、市场、交易等内容有深刻的认识；二是从业人员要有扎实的金融投资专业知识，对金融市场、金融交易、金融风险收益有深刻的理解和分析。

19.56克拉缅甸无烧矢车菊蓝蓝宝石戒指

Luxury Jewelry Investment and Appreciation

Luxury Jewelry Investment Advisory Services refer to specialized asset allocation consulting services provided by providers of such services to clients.

Luxury jewelry investment services have the characteristics of consulting, only to provide customers with professional advice on luxury jewelry investment, but not involving the specific operation of customer asset allocation. The customers have the final decision-making power of financial resources. Luxury jewelry investment advisory service has professional characteristics. First, practitioners should have solid professional knowledge of luxury jewelry and have a deep understanding of the types, markets and transactions of luxury jewelry. Second, practitioners should have solid professional knowledge of financial investment, and have a deep understanding and analysis of financial markets, financial transactions, and financial risks and returns.

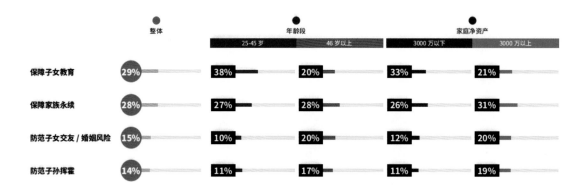

家庭现金流管理的目的
资料来源：中信保诚人寿（传家）胡润百富2023中国高净值家庭现金流管理报告
Objectives of Family Cash Flow Management
Source: CITIC-Prudential (Legacy) Hurun Hundred Sayings 2023 China High-Net-Worth Family Cash Flow Management Report

二、名贵珠宝投资顾问服务中的客户画像
Portraits of Clients in Luxury Jewelry Investment Advisory Services

（一）客户信息收集
Customer Information Collection

在名贵珠宝投资顾问服务中，要对目标客户信息进行精准分析，以提升投资顾问的收益、降低风险。要了解目标客户的财务信息，包括客户的收支状况、财务目标、财务规划、风险特征、投资偏好、投资知识水平和决策模式等。要了解目标客户的非财务信息，包括目标客户的年龄、健康状况、人际交往、社会地位、家庭关系、兴趣爱好、未来生活方式等。

在名贵珠宝投资顾问服务过程中，可通过与客户沟通交流，了解客户个人的社会信息、财务信息和投资决策信息，并分析宏观经济中的经济金融数据，以更好地建立客户信息数据库。一是建立客户财务收入数据库，了解客户个人的资产负债情况，通过现金流量表分析客户的现金收入和现金支出状况，找到客户在名贵珠宝投资上存在的潜在问题，帮助客户更好地利用财务资源。二是建立客户财务支出数据库，预测客户未来的收入情况，并根据宏观经济金融状况对客户未来收入情况进行合理估算，包括估算客户工资、股票、债券、存款、租金等常规性收入在未来的涨跌情况，估算客户临时性收入在未来的涨跌情况，预测客户未来的基本生活支出状况以及高品质生活和奢侈消费支出状况，从而估计客户未来在名贵珠宝投资上的潜力和偏好。

In the luxury jewelry investment advisory service, it is necessary to accurately analyze the target customer information to improve the return of investment advisers and reduce risks. To understand the target customer's financial information, including the customer's income and expenditure status, financial objectives, financial planning, risk characteristics, investment preferences, investment knowledge level and decision-making mode. To understand the target customer's non-financial information, including the target customer's age, health, interpersonal communication, social status, family relations, interests, future lifestyle, etc.

In the process of luxury jewelry investment consulting service, through communication with customers, we can understand customers' personal social information, financial information and investment decision information, and analyze economic and financial data in the macro economy to better establish customer information database. The first is to establish a customer financial income database, to understand the customer's personal assets and liabilities, to analyze the cash income and cash expenditure of the customer through the cash flow statement, and to find the potential problems of the customer's investment in luxury jewelry, and to help the customer make better use of financial resources. The second is to establish the customer's financial expenditure database, to predict the customer's future income, and to make reasonable estimation of the customer's future income according to the macroeconomic and financial conditions, including estimating the future rise and fall of the customer's regular income such as salary, stock, bond, deposit and rent, and estimate the future rise and fall of the customer's temporary income. To predict the customer's future basic living expenses as well as high quality living and luxury consumption expenses, so as to estimate the customer's future potential and preference for luxury jewelry investment.

（二）客户风险特征分析和投资特征分析
Customer Risk Characteristics Analysis and Investment Characteristics Analysis

在名贵珠宝投资顾问服务中，要高度关注目标客户的风险特征和投资特征。就风险特征而言，不同的目标客户对风险的感知完全不同，要分析客户在主观上对风险的态度即风险偏好，分析客户对风险的认知程度，分析客户对风险的承受能力即风险厌恶型、风险偏爱型和风险中立型，这些都会影响客户对投资型名贵珠宝和投机型名贵珠宝的投资选择。就理财特征而言，要关注目标客户的投资渠道偏好，关注客户的投资知识结构，关注客户的生活方式和个人性格，这些会影响客户对名贵珠宝投资产品、投资渠道和投资方式的抉择。

In the luxury jewelry investment advisory service, we should pay great attention to the risk characteristics and investment characteristics of the target customers. In terms of risk characteristics, different target customers have completely different perceptions of risk. It is necessary to analyze customers' subjectively attitude towards risk (risk preference), their cognitive degree of risk, and their risk tolerance (risk aversion, risk preference and risk neutrality). These will affect the customer's investment choice of investment type or speculation type of luxury jewelry. In terms of financial characteristics, it is necessary to pay attention to the investment channel preferences of target customers, to pay attention to the investment knowledge structure of customers, to pay attention to the lifestyle and personal personality of customers, which will affect the choice of customers on luxury jewelry investment products, investment channels and investment methods.

（三）客户名贵珠宝理财需求和投资需求分析
Analysis of Customers' Financing Needs and Investment Needs for Luxury Jewelry

在名贵珠宝投资顾问服务中，要关注目标客户的理财需求和投资需求。从投资理财的时间周期看，客户的投资目标通常包括存款、购车、日常消费等短期目标，包括教育、买房、投资等中期目标，包括养老、资产保值增值、财富传承等长期目标。因此，名贵珠宝投资顾问服务要解决客户投资目标和投资风险之间的矛盾，通过名贵珠宝投资优化客户资产配置，提升客户资产的购买能力、投资能力、抗风险能力和传承能力，充分保障高净值客户的品质生活、家族财富安全、家族财富传承。

In the luxury jewelry investment advisory service, we should pay attention to the financing needs and investment needs of target customers. From the perspective of the time cycle of investment and financial management, customers' investment goals usually include short-term goals such as deposit, car purchase and daily consumption; medium-term goals such as education, house purchase and investment; and long-term goals such as old-age care, asset preservation and appreciation, and wealth inheritance. Therefore, luxury jewelry investment advisory services should solve the contradiction between customers' investment objectives and investment risks, and optimize customers' asset allocation through luxury jewelry investment. It can also improve customers' asset purchasing ability, investment ability, risk resistance ability and inheritance ability, and fully protect the quality of life, family wealth security and family wealth inheritance of high net worth customers.

三、名贵珠宝投资顾问服务中的财务规划
Financial Planning in Luxury Jewelry Investment Advisory Services

名贵珠宝投资顾问服务要在全面了解客户的基础上，为客户提供精准的财务规划，通过现金流管理实现客户现金结余的合理安排，通过投资规划实现客户资产保值增值和风险转移，为客户打造一个安全、增值的财务规划体系。

On the basis of comprehensive understanding of customers, luxury jewelry investment advisory services should provide customers with accurate financial planning, achieve reasonable arrangement of cash balances through cash flow management, maintain and increase the value of customers' assets and transfer risks through investment planning, and create a safe and value-added financial planning system for cus-

（一）名贵珠宝投资顾问的财务功能
Financial Functions of Luxury Jewelry Investment Advisers

名贵珠宝投资顾问对高净值家庭而言，具有非常重要的意义，能够提升家庭资产的购买能力、投资能力、偿付能力、抗风险能力和传承能力。稳定、持续、正向的财务现金流管理，能够有效保障高净值家庭的高品质生活、资产增值保值和财富传承。高品质生活和资产保值升值是高净值家庭的中短期目标，而家族财富传承是高净值家庭的远期目标。

Luxury jewelry investment advisor is of great significance for high net value families, which can improve the purchasing ability, investment ability, solvency, anti-risk ability and inheritance ability of family assets. Stable, continuous and positive financial cash flow management can effectively guarantee the high quality of life, asset appreciation and preservation and wealth inheritance of high net value families. High quality of life and asset preservation and appreciation are the short- and medium-term goals of high net value families, while family wealth inheritance is the long-term goal of high net value families.

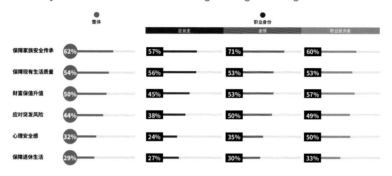

家庭现金流管理的目的
资料来源：中信保诚人寿（传家）胡润百富2023中国高净值家庭现金流管理报告
Objectives of Family Cash Flow Management
Source: CITIC-Prudential (Legacy) Hurun Hundred Sayings 2023 China High-Net-Worth Family Cash Flow Management Report

（二）名贵珠宝投资顾问的投资规划
Investment Planning of Luxury Jewelry Investment Consultants

在为客户制定名贵珠宝投资规划的过程中，要有一系列完整的投资规划。要根据客户的投资目标和可投资财富数量，参照客户的风险偏好，确定采取投资型的名贵珠宝配置策略还是投机型的名贵珠宝配置策略，合理安排名贵珠宝的配置比例，并根据投资业绩对名贵珠宝投资组合进行必要的调整。一是明确客户的投资目标，尤其是高净值家庭在高品质生活、奢侈消费、财富保值升值和家族财富传承等方面的需求。二是明确客户的风险承受能力，根据客户的年龄、资产状况和风险测试，判断客户对风险承受能力和投资收益率的要求，确定符合客户需求的组合投资策略，从而配置不同类型的名贵珠宝资产。三是实施和监控投资计划，密切关注名贵珠宝资产的涨跌情况，看是否达到了预期投资目标，并紧密跟踪客户投资预期，根据客户要求和市场变动状况，调整投资目标和策略，增加或者减持名贵珠宝资产。

In the process of formulating investment planning for customers, there should be a series of complete investment planning. According to the customer's investment objectives and investable wealth, and with reference to the customer's risk preference, it is necessary to determine whether to adopt the investment type of luxury jewelry allocation strategy or the speculation type of luxury jewelry allocation strategy, and rationally arrange the allocation proportion of luxury jewelry to make necessary adjustments to the luxury jewelry portfolio according to the investment performance. The first is to clarify the investment objectives of customers, especially the needs of high net value families in terms of high-quality life, luxury consumption, wealth preservation and appreciation, and family wealth inheritance. The second is to clarify the customer's risk tolerance, judge the customer's requirements on risk tolerance and investment return according to the customer's age, asset status and risk test, and determine the portfolio investment strategy that meets the customer's needs, so as to allocate different types of luxury jewelry assets. The third is to implement and monitor the investment plan, pay close attention to the rise and fall of luxury jewelry assets to see whether the expected investment target has been reached, and closely track the investment expectations of customers, adjust the investment objectives and strategies according to customer requirements and market changes, and increase or reduce the holding of luxury jewelry assets.

Luxury Jewelry Appreciation and Investment

第七章
07 Chapter Seven

附录：
名贵珠宝投资资源
Appendix: Luxury Jewelry
Investment Resources

一、主要的名贵珠宝评估和鉴定机构名录
List of Major Luxury Jewelry Evaluation and Appraisal Institutions

名贵珠宝鉴定机构 Luxury Jewelry Appraisal Agency	实验室分布 Laboratory Distribution	介绍 Introduction	网站 Website
古柏林 Gubelin	总部：瑞士卢塞恩 全球鉴定所： 中国香港名贵珠宝鉴定所 美国纽约名贵珠宝鉴定所 Headquarters: Lucerne, Switzerland Global appraisal Institute: China Hong Kong luxury jewelry appraisal Institute the United States New York luxury jewelry appraisal Institute	古柏林名贵珠宝实验室建立于1923年。是目前世界上最为著名的名贵珠宝实验室之一，同时也是世界上彩色名贵珠宝鉴定最权威的机构之一。佳士得和苏富比对其拍卖会上的重要名贵珠宝均要求出具古柏林的名贵珠宝鉴定证书，可见古柏林证书的权威性。 The Gubelin Luxury jewelry Laboratory was established in 1923. At present, it is one of the most famous luxury jewelry laboratories in the world, and it is also one of the most authoritative institutions for the identification of colored luxury jewelry in the world. Christie's and Sotheby's on their auction of important luxury jewelry are required to issue Gubelin's luxury jewelry identification certificate, which shows the authority of Gubelin's certificate.	http://www.gubelingemlab.ch
瑞士珠宝研究院 SSEF	总部：瑞士巴塞尔 Headquarters: Basel Switzerland	全称Swiss Gemmological Institute，简称SSEF，成立于1974年，瑞士的三大鉴定机构之一，是一个非营利性的独立科学研究机构。 由于鉴定的专业性和历史性深受世界知名拍卖行、皇族、珠宝商、收藏家的信赖。 Swiss Gemmological Institute, referred to as SSEF, was founded in 1974, one of the three major identification institutions in Switzerland, is a non-profit independent scientific research institution. Because of the professionalism and history, it earned deep trust from the world's famous auction houses, royal families, jewelers, and collectors.	https://www.ssef.ch

名贵珠宝鉴定机构 Luxury Jewelry Appraisal Agency	实验室分布 Laboratory Distribution	介绍 Introduction	网站 Website
GRS	总部： 瑞士 全球办事处： 美国 法国 斯里兰卡 泰国 中国香港 Headquarters: Switzerland Global offices: United States France Sri Lanka Thailand Hong Kong China	创立于1996年，位于瑞士，全称Gem Research Swiss Lab。瑞士三大鉴定机构之一，是Dr. A. Peretti创立的一家私人名贵珠宝实验室，常年为全世界最顶尖的珠宝拍卖公司提供商业鉴定服务，是彩色名贵珠宝界名声最响、市场占有率最高的证书。 Founded in 1996 and located in Switzerland, the full name is Gem Research Swiss Lab.One of the three major appraisal institutions in Switzerland, is a private luxury jewelry laboratory founded by Dr. A. Peretti, which provides commercial appraisal services for the world's top jewelry auction companies all year round, and is the certificate with the highest reputation and market share in the color luxury jewelry industry.	https://www.gemresearch.ch
国际名贵 珠宝学院 IGI	总部： 比利时安特卫普 全球办事处： 纽约、香港、孟买、 曼谷、东京、 迪拜、特拉维夫、 多伦多、洛杉矶、 加尔各答、新德里、德里久尔、 斋浦尔、苏拉特、金奈、 阿默达巴德和海得拉巴 Headquarters: Antwerp, Belgium Global offices: New York, Hong Kong, Mumbai, Bangkok, Tokyo, Dubai, Tel Aviv, Toronto, Los Angeles, Kolkata, New Delhi, Delhi, Jaipur, Surat, Chennai, Ahmedabad and Hyderabad	是最古老的名贵珠宝学院，世界三大权威鉴定机构之一，也是全球最大的独立鉴定珠宝首饰的实验室，在各个国家的大型交易中心都设有实验室。 It is the oldest luxury jewelry school, one of the world's three authoritative appraisal institutions, and the world's largest independent appraisal jewelry laboratory, with laboratories in large trading centers in various countries.	http://www.igiworldwide.com

名贵珠宝鉴定机构 Luxury Jewelry Appraisal Agency	实验室分布 Laboratory Distribution	介绍 Introduction	网站 Website
英国名贵珠宝协会和名贵珠宝检测实验室 FGA/Gem-A	英国 UK	成立于1908年，全称Gemmological Association & Gem Testing Laboratory of Great Britain。世界上最老的名贵珠宝协会，1931年成为独立的机构。1990年和Gem Testing Laboratory of Great Britain形成合作后，在2000年启用了新缩写Gem-A。FGA的鉴定机构非常出名，但证书在市场很少，反而是以考试出名。通过FGA考试的学生被授予名贵珠宝学证书，随后他们可以申请英国名贵珠宝协会会员资格，即Fellow of the Gemmological Association (FGA)。 It was founded in 1908 as the Gemmological Association & Gem Testing Laboratory of Great Britain.The oldest fine jewelry association in the world, it became an independent institution in 1931. After forming a partnership with Gem Testing Laboratory of Great Britain in 1990, the new acronym GEM-A was adopted in 2000.FGA's certification institution is very famous, but the certificate is rarely in the market, but it is famous for the exam. Students who pass the FGA are awarded a Certificate in Fine Jewelry and can then apply for membership of the British Fine Jewelry Association, the Fellow of the Gemmological Association (FGA).	https://gem-a.com
美国名贵珠宝学院 GIA	美国加州 California, USA	成立于1931年，位于美国加州的Carlsbad，全称Gemological Institute of America。买钻石不可能绕过的名字，GIA提出了钻石的4C标准，并把这个标准推广向全世界。在钻石鉴定方面，GIA是全球公认最权威的。除了鉴定钻石，GIA也鉴定名贵珠宝。 It was founded in 1931, located in Carlsbad, California, the full name of the Gemological Institute of America.Buying diamonds is impossible to bypass this name. GIA proposed the diamond 4C standard, and promote this standard to the world. In the field of diamond identification, GIA is recognized as the most authoritative authority in the world, in addition to the identification of diamonds, GIA also identifies luxury jewelry.	http://www.gia.edu

名贵珠宝鉴定机构 Luxury Jewelry Appraisal Agency	实验室分布 Laboratory Distribution	介绍 Introduction	网站 Website
美国名贵珠宝学院 AGL	美国纽约 New York, USA	成立于1977年，位于美国纽约，全称American Gemological Laboratories。AGL是公认的有色名贵珠宝鉴定的权威机构之一，是美国第一家推出名贵珠宝原产地鉴定意见的名贵珠宝实验室，也是第一个采用量化的方法对彩色名贵珠宝品质的各个指标进行分析的机构。业内人士非常认可，也是苏富比拍卖认可的彩色名贵珠宝鉴定机构。 Founded in 1977 and located in New York, USA, the full name is American Gemological Laboratories.AGL is recognized as one of the authoritative institutions for the identification of colored luxury jewelry. It is the first luxury jewelry laboratory in the United States to introduce the identification opinions of luxury jewelry origin, and is also the first institution to adopt a quantitative method to analyze the various indicators of colored luxury jewelry quality. Highly recognized by the industry, it is also a colored luxury jewelry appraisal institution recognized by Sotheby's auction.	http://aglgemlab.com
吉尔德 GUILD	总部： 美国洛杉矶 全球实验室： 香港、深圳设有实验室 Headquarters: LosAngeles, USA Global laboratories: There are laboratories in Hong Kong and Shenzhen	成立于1980年，全称Guild Gem Laboratories。是一家专注于彩色名贵珠宝检测的机构，由美国名贵珠宝学家Mr.Charles Carmona创立。在彩色名贵珠宝鉴定、分级、产地研究方面具有丰富的经验，在祖母绿方面具有一定的权威性。因为在深圳开了实验室，现在中国使用得比较多。 Founded in 1980 as Guild Gem Laboratories.It is an institution focused on the detection of colored luxury jewelry, founded by the American precious jeweler Mr.Charles Carmona. It has rich experience in identification, classification and origin research of colored luxury jewelry, and has certain authority in emerald.As the laboratory was opened in Shenzhen, it is now used more in China.	http://www.guildgemlab.com

名贵珠宝鉴定机构 Luxury Jewelry Appraisal Agency	实验室分布 Laboratory Distribution	介绍 Introduction	网站 Website
亚洲名贵珠宝学院 AIGS	泰国曼谷 Bangkok, Thailand	于1978年由著名华裔商人何荣光先生在泰国曼谷成立,是首次在世界上研究出红、蓝宝石的分级法、分级标准的实验室,该分级体系至今在世界范围内被广泛使用,在红、蓝宝石的优化处理和产地鉴定方面的权威性,也得到世界彩色名贵珠宝界的一致认可。40年的经验让AIGS在对产地鉴定上处于世界领先的位置。另外,AIGS还提供翡翠鉴定报告,在翡翠分级上也有着普遍认可的权威性,不仅能鉴定翡翠的天然性、是否经历过漂白、染色处理,还能对翡翠的质量进行分级。 It was established in 1978 by the famous Chinese businessman Mr. He Rongguang in Bangkok, Thailand. It is the first laboratory in the world to study the grading method and grading standards of red and sapphire. The grading system has been widely used in the world, in the optimization of red and sapphire treatment and origin identification authority, but also the world's colored luxury jewelry industry unanimously recognized. Over 40 years of experience, AIGS is a world leader in origin identification.In addition, AIGS also provides jade identification report, which also has a generally recognized authority in the classification of jade, which can not only identify the natural nature of jade, whether it has bein bleached and dyed, but also grade the quality of jade.	http://aglgemlab.com
斯里兰卡 名贵珠宝实验室 EGL	斯里兰卡 Sri Lanka	全称Emteem Gem Lab,是斯里兰卡当地的一个鉴定实验室,位于Beruwela,主要用于鉴定蓝宝石、红宝石、金绿猫眼等彩色名贵珠宝,可以对红蓝宝是否经过加热处理作出说明。 The full name is Emteem Gem Lab. It is a local identification laboratory in Sri Lanka, located in Beruwela, mainly used to identify sapphire, ruby, gold and green cat's eye and other color precious jewelry. It can indicate whether the red and blue treasure has been heated.	http://emteemgemlab.lk

名贵珠宝鉴定机构 Luxury Jewelry Appraisal Agency	实验室分布 Laboratory Distribution	介绍 Introduction	网站 Website
国家珠宝玉石质量监督检验中心 NGTC	总部：北京 Headquarters: Beijing	NGTC全称为National Gemological Centre，即国家珠宝玉石质量监督检验中心，隶属于国土资源部珠宝玉石首饰管理中心，长期以来一直致力于珠宝职业教育和国家珠宝玉石标准的推广和普及工作。NGTC同时也是美国名贵珠宝学院(GIA)在中国内地唯一的合作伙伴。 NGTC, the full name of the National Gemological Training Center, that is, the National Jewelry and Jade Quality Supervision and Inspection Center. It belongs to the jewelry and jade Jewelry Management Center of the Ministry of Land and Resources, and has long been committed to jewelry vocational education, the promotion and popularization of national jewelry and jade standards. NGTC is also the sole partner of the Gemologic Institute of America (GIA) in Chinese mainland.	http://aglgemlab.com

二、名贵珠宝展览、交易会和拍卖信息
Luxury Jewelry Exhibition, Trade Fair and Auction Information

(一) 名贵珠宝展览
Luxury Jewelry Exhibition

国家 Country	展览名称 Name
中国 China	中国国际珠宝展 China International Jewelry Fair
	中国香港国际珠宝首饰展览会 China Hong Kong International Jewelry Fair
	中国香港国际钻石、名贵宝石及珍珠展览会 China Hong Kong International Diamond, Luxury jewelry and Pearl Exhibition
	深圳国际珠宝展览会 Shenzhen International Jewelry Fair

美国 American	美国图森国际矿 Tucson International Mineral and Jewelry
	美国拉斯维加斯国际珠宝展览会(JCK珠宝展) Las Vegas International Jewelry Exhibition(JCK Jewelry Exhibition) Gem Fair
	美国纽约的J·A国际珠宝展览会 J·A International Jewelry Show in New York, USA
瑞士 Switzerland	巴塞尔国际钟表珠宝展 Baselworld
	瑞士日内瓦珠宝展览会 Geneva Jewelry Fair, Switzerland Gem Fair
意大利 Italy	意大利维琴察珠宝展览会 Vicenza Jewelry Fair, Italy
德国 Germany	德国慕尼黑珠宝钟表展览会 Munich Jewelry and Watch Fair in Germany
英国 U.K.	英国伯明翰国际珠宝钟表展览会 Birmingham International Jewelry and Watch Fair
	英国伦敦国际珠宝展览会 London International Jewelry Fair
法国 France	法国巴黎珠宝钟表展览会 Paris Jewelry and Watch Fair
日本 Japan	日本东京珠宝展览会 Tokyo Jewelry Exhibition, Japan
	日本神户国际珠宝展览会 Kobe International Jewelry Exhibition, Japan
泰国 Thailand	泰国曼谷国际珠宝玉石首饰展览会 Bangkok International Jewelry and Jade Jewelry Exhibition, Thailand
斯里兰卡 Sri Lanka	斯里兰卡科伦坡珠宝名贵宝石展览会 Sri Lanka Colombo Jewelry and Precious Gem Exhibition
新加坡 Singapore	新加坡珠宝展览会 Singapore Jewelry Fair
印度 India	印度海德拉巴珠宝、珍珠及名贵宝石展览会 Jewelry, Pearls and Precious Gems Exhibition in Hyderabad, India
	印度孟买国际珠宝展 Mumbai International Jewelry Exhibition, India

(二)珠宝交易所
Jewelry Trading Institution

国家 Country	交易所名称 Name
中国 China	上海钻石交易所 Shanghai Diamond Trading
美国 USA	纽约曼哈顿中城 47 街 47th Street, Midtown Manhattan, New
英国 UK	伦敦霍尔本哈顿花园 Hatton Gardens, Holborn, London
斯里兰卡 Sri Lanka	科伦坡贝鲁沃勒 Beruwala, Colombo

(三)拍卖
Auction House

拍卖行名称 Auction House	简介 Introduction
苏富比 Sotheby's	1744年创立于伦敦，是世界上最古老的拍卖行，也是规模最大的国际知名艺术拍卖行。世界拍卖史上最昂贵的十大珠宝，很多都来自苏富比。 Founded in London in 1744, Sotheby's is the world's oldest auction house and the largest internationally renowned art auction house. Many of the top 10 most expensive jewels ever auctioned in the world come from Sotheby.
佳士得 CHRISTIE'S	佳士得创建于1766年，世界著名艺术品拍卖行之一。拍品汇集了来自全球各地的珍罕艺术品、名表、珠宝首饰、汽车和名酒等精品。中国香港佳士得被选作为亚洲拍卖的翘楚，每年在5月和11月举行两次拍卖活动。 Founded in 1766, Christie's is one of the world's leading art auction houses. The auction brings together rare works of art, watches, jewelry, cars and wines from all over the world. Christie's Hong Kong, China is selected as a leader in Asian auctions, holding two sales each year in May and November.

拍卖行名称 Auction House	简介 Introduction
邦瀚斯 Bonhams	邦瀚斯拍卖行是全球历史最悠久、规模最大的拍卖行之一，创立于1793年，该拍卖行设有57个分类的专研珍藏部门，专业搜寻及评鉴各国各门类的珍藏。 Bonhams is one of the oldest and largest auction houses in the world, founded in 1793, the auction house has 57 categories of specialist collection departments, specializing in the search and evaluation of national collections.
嘉德拍卖 China Guardian (HK) Auctions	嘉德拍卖成立于1993年5月，是以经营中国文物艺术品为主的综合型拍卖公司，每年都会定期举办春、秋两次大型拍卖会，同时还会举办线上拍卖会。 Founded in May 1993, Guardian Auction is a comprehensive auction company specializing in Chinese cultural relics and works of art. It regularly holds two large-scale auctions in spring and autumn every year, as well as online auctions.
保利拍卖 Beijing Poly International Auction	保利拍卖是一家成立于2005年的隶属于保利文化集团股份有限公司的大型国有控股拍卖公司。 Established in 2005, Poly Auction is a large state-owned holding auction company affiliated to Poly Culture Group Co., LTD.

图片来源
SOURCE OF THE PICTURE

BESTFORU · 百睿珠宝

金百灵珠宝
陈艺明
顽石

切个石头
Frantiques · 梵醒客古董珠宝
邂逅巴黎 汪小仙
韦小陶

G.GEM夏至
许泽坚
翠心珠宝
卡凡珠宝

佳士得拍卖
苏富比拍卖
邦瀚斯拍卖
北京匡时拍卖

IWO珠宝
诚好珠宝
SUNTION尚宸珠宝
Thailand Coka Gem

彩臻珠宝
启航珠宝
GULID宝石实验室
GIA美国宝石研究院

三、其他珠宝鉴赏
Appreciation of Other Jewelries

Aquamarine

25.30克拉超级圣玛丽亚海蓝宝
25.30 ct Exceptional Santa Maria Color Aquamarine

21.91克拉圣玛丽亚海蓝宝戒指
21.91ct Santa Maria Color Aquamarine Ring

1 海蓝宝 Aquamarine

海蓝宝是一种非常珍贵的宝石，与祖母绿是同一个品种。优质海蓝宝的颜色为天蓝色至海蓝色，最好的颜色评级为"圣玛丽亚色－超级圣玛丽亚色"。我们在购买的时候，如果希望保值，尽量选择颜色浓郁、净度好、火彩好的，大克拉的高品质海蓝宝很受国际珠宝大牌的欢迎。

Aquamarine is a very precious jewelry, and it is the same variety as the emerald. Premium aquamarine colors range from sky blue to sea blue, and the best color rating is "Santa Maria - Super Santa Maria." When we do purchasing, if we want to preserve value, try to choose rich color, good clarity, good fire color. The large carat high-quality aquamarine is very popular with international jewelry brands.

Chaffre

沙弗莱戒指
Tsavorite Ring

沙弗莱
Tsavorite

2 沙弗莱 Chaffre

沙弗莱是石榴石的一种，颜色很浓郁，看起来很像祖母绿，有一种绿野仙踪的感觉。我们在购买的时候，如果希望保值，尽量选择绿色调浓郁，不含杂色，火彩好和切割好的。10克拉左右的高品质沙弗莱很稀有，很受国际珠宝大牌的欢迎。

Chaffre is a type of garnet that is very rich in color and looks very much like emeralds, with a sense of the Wizard of Oz. When we do purchasing, if we want to preserve value, try to choose a strong green tone, no miscellaneous color, fire color and cut well. High quality Chafferais of around 10 carats are very rare and very popular with international jewelry brands.

Fanta Stone 芬达石

3 芬达石
Fanta Stone

芬达石 Spessartine

芬达石戒指 Spessartine Ring

芬达石也是石榴石的一种，是最有活力的一种宝石，具有像芬达汽水一样的橙色调，被称为"宝石界的小太阳"。我们在购买的时候，如果希望保值，尽量选择颜色浓郁、切割好、火彩好的。大克拉的芬达石，也非常受国际珠宝大牌的欢迎。

Fanta Stone is also a type of garnet, is one of the most dynamic gemstones, with an orange tone like Fanta soda, known as the "little sun of the gem world." When we do purchasing, if we want to preserve value, try to choose a strong color, cut well, and fire color. The large carat Fanta stone is also very popular with international jewelry brands.

芬达石

Green Garnet 翠榴石

4 翠榴石
Green Garnet

翠榴石 Demantoid

11克拉翠榴石戒指 11ct Demantoid Ring

翠榴石是石榴石中非常稀有的宝石，它具有半金刚光泽和高色散，非常闪亮，是世界上最闪耀的宝石之一，有些极品翠榴石，比钻石更闪耀。它最大的特征是内部有马尾丝状的包裹体，非常醒目，但是翠榴石长不大，3克拉的翠榴石都是精品了。俄罗斯乌拉尔山脉的翠榴石非常受宝石收藏者的欢迎。

Green Garnet is a very rare gemstone in garnet, it has a semi-diamond luster and high dispersion, very shiny, is one of the most shining gemstones in the world. Some of the best green garnet, more shining than diamonds. Its biggest feature is the inclusion of horsetail inside, very eye-catching, but the green garnet is not big, 3 carats of green garnet are considered very fine. The green garnet from Russia's Ural Mountains is very popular among gem collectors.

翠榴石

Apprecation of Other Jewelries 5/6 其他宝石鉴赏

Tour ma line 碧玺

5 碧玺 Tourmaline

碧玺就是电气石,它的品类非常多,颜色也非常多。比较保值的是阿富汗碧玺。拉贡(Lagoon)碧玺,它有电光质感,颜色很纯,火彩很好,其中蓝色调和粉色调都非常受欢迎,很多国际珠宝大牌都会用它来做设计。

Tourmaline

categories are very various, the color is also very various. One that holds its value is the Afghan tourmaline. Lagon tourmaline has an electric texture, its color is very pure, the fire color is very good, which blue and pink tones are very popular, many international jewelry brands will use it to design.

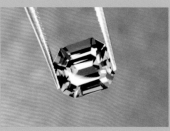

拉贡碧玺戒指
Lagoon Tourmaline Ring

拉贡碧玺
Lagoon Tourmaline

碧玺

Chry sob eryl 金绿宝石

6 金绿宝石 Chrysoberyl

金绿宝石,它有一个非常稀有的品种叫钒金绿,绿色调非常迷人,火彩非常好,长得最像绿钻石。钒金绿很受欧美珠宝人的欢迎,价格也非常贵。金绿宝石还有一个很受欢迎的品种,像荧光笔一样的柠檬黄色,非常招人喜欢。

Chrysoberyl,

it has a very rare variety called vanadium gold green, whose green tone is very charming, fire color is very good, and looks most like green diamond. Vanadium gold green is very popular with European and American jewelers, and the price is also very expensive. Chrysoberyl has another very popular variety, a lemon yellow color like a highlighter, which is very popular.

5克拉 钒金绿
5ct Vanadium Chrysoberyl

钒金绿戒指
Vanadium Chrysoberyl Ring

金绿宝石

其他宝石鉴赏 Apprecation of Other Jewelries 7/8

Tanzan坦桑石ite

7 坦桑石
Tanzanite

7

坦桑石是坦桑尼亚的国石,很浓郁的蓝色调,长很像蓝宝石。但是坦桑石大多数都会带有一点紫色调,但并不影响它整体的美观。

Tanzanite is the national stone of Tanzania, with very rich blue color, looking very like sapphire. However, most tanzanite will have a little purple tone, but it does not affect its overall beauty.

坦桑石
Tanzanite

坦桑石项链
Tanzanite Necklace

坦桑石

Per id橄榄石ot

8

8 橄榄石
Peridot

橄榄石是一种非常宁静的宝石,它的颜色主色调为绿色,但是大多数宝石带一点黄色调。我们在购买的时候,如果考虑保值,尽量买黄色调少一点的缅甸产的橄榄石,绿色调非常浓郁,很像祖母绿。

Peridot is a very peaceful gemstone, its color is mainly green, but most gemstones have a slight yellow tinge. When we do purchasing, if we consider value preservation, try to buy Burmese peridot with less yellow tone, and with strong green tone which like emerald.

橄榄石
Peridot

橄榄石戒指
Peridot Ring

橄榄石

图书在版编目(CIP)数据

名贵珠宝鉴赏与投资：汉英对照 / 陈小雨主编.
上海：上海社会科学院出版社，2024. -- ISBN 978-7
-5520-4468-3

Ⅰ．TS933.21：F768.7

中国国家版本馆CIP数据核字第2024J97E30号

名贵珠宝鉴赏与投资——便携式资产配置

主　　编：陈小雨
副 主 编：田　卫
统　　筹：邢萱萱
责任编辑：霍　覃
封面设计：刘春燕　江姚君　魏　昊
出版发行：上海社会科学院出版社
　　　　　上海顺昌路622号　邮编200025
　　　　　电话总机 021-63315947　销售热线 021-53063735
　　　　　https://cbs.sass.org.cn　E-mail：sassp@sassp.cn
排　　版：刘春燕　江姚君　魏　昊
印　　刷：上海雅昌艺术印刷有限公司
开　　本：889毫米×1194毫米　1/16
印　　张：19
插　　页：1
字　　数：500千
版　　次：2024年9月第1版　2025年5月第2次印刷

ISBN 978-7-5520-4468-3/TS·021　　　　　　　定价：298.00元

版权所有　翻印必究